PRINCIPLES AND STANDARDS FOR THE DISPOSAL OF LONG-LIVED RADIOACTIVE WASTES

Waste Management Series

1. Waste Materials in Construction. The Science and Engineering of Recycling for Environmental Protection
   Edited by G.R. Woolley, J.J.J.M. Goumans and P.J. Wainwright

2. Geological Disposal of Radioactive Wastes and Natural Analogues. Lessons from Nature and Archeology.
   By W. Miller, R. Alexander, N. Chapman, I. McKinley and J. Smellie

Other relevant titles from Elsevier/Pergamon

Municipal Solid Waste Incinerator Residues 1997 By: A.J. Chandler, T.T. Eighmy, J. Hartlén, O. Hjelmar, D.S. Kosson, S.E. Sawell, H.A. van der Sloot, J. Vehlow

Harmonization of Leaching/Extraction Tests 1997 Edited by H.A. van der Sloot, L. Heasman, Ph. Quevauviller

Waste Materials in Construction: Putting Theory into Practice 1997 Edited by J.J.J.M. Goumans, G.J. Senden, H.A. van der Sloot

Waste Management Series, Volume 3

PRINCIPLES AND STANDARDS FOR THE DISPOSAL OF LONG-LIVED RADIOACTIVE WASTES

Neil Chapman
Department of Civil and Structural Engineering, University of Sheffield, UK

Charles McCombie
McCombie Consulting, Switzerland

ELSEVIER

Amsterdam – Boston – Heidelberg – London – New York – Oxford
Paris – San Diego – San Francisco – Singapore – Sydney – Tokyo

ELSEVIER B.V.
Sara Burgerhartstraat 25
P.O. Box 211, 1000 AE
Amsterdam, The Netherlands

ELSEVIER Inc.
525 B Street
Suite 1900, San Diego
CA 92101-4495, USA

ELSEVIER Ltd
The Boulevard
Langford Lane, Kidlington,
Oxford OX5 1GB, UK

ELSEVIER Ltd
84 Theobalds Road
London WC1X 8RR
UK

First edition 2003
Second Impression 2004

Library of Congress Cataloging in Publication Data
A catalog record is available from the Library of Congress.

British Library Cataloguing in Publication Data
A catalogue record is available from the British Library.

ISBN: 0-08-044192-0

♾ The paper used in this publication meets the requirements of ANSI/NISO Z39.48-1992 (Permanence of Paper).
Printed in The Netherlands.

To Lotti and Liz

Preface

This book originated in a project undertaken to provide advice to the Japanese nuclear industry on establishing principles and setting standards for the disposal of radioactive wastes. The study was designed to provide input for the reorganisation in Japan in 2000 of the policies and organisational structures for long-term management of such wastes. It was clear, however, that many of the issues being considered had wider interest outside the strictly Japanese context, and even outside that of radioactive waste disposal. The issue of how a society best fulfils its responsibility for protecting people and the environment from the hazards of radioactive materials has to be looked at in the wider context of management of toxic materials and this in turn has to be seen against the broader background of how health and environmental policies are determined.

Consequently, we decided to revise and extend the original work and offer it to a wider readership as input to the current international debate on a range of complex issues associated with technical undertakings that can have societal effects extending far into the future. These issues include the protection of future generations, social equity considerations in current and future societies, the feasibility of making quantitative predictions about the future and the challenge of making decisions in the face of social and scientific uncertainties. These topics are important, and they should all be considered in decisions on the deployment of societal resources and in the management of technological undertakings that can affect peoples' health and well being (today or in the future).

The issues addressed are relevant for the increasing numbers of modern technologies that must balance potentially beneficial and detrimental effects extending into the distant future, far beyond the timescales of direct concern to those developing and deploying the technologies. Radiation protection in the nuclear field, the general topic addressed in this review, does not present the biggest hazard amongst all such technologies. Waste disposal, the specific topic, does not provide the greatest challenge in radiation protection. Why, then, do we use safety in waste disposal as the focus? The answer is that much effort has been devoted by the waste management community to debating the relevant issues and to constructing a coherent set of principles and standards. The extensive, perhaps even disproportionate, resources which it has been possible to devote to these developments are due in large measure to the fact that a major global industry, that of nuclear power production, has recognised that demonstrating environmentally acceptable waste disposal is an essential prerequisite for its continuance. Accordingly, pioneering work has been done on the scientific and societal questions concerning principles and standards. This work, however, is not familiar, even to many of those in the field, and is virtually unrecognised in other scientific areas or by the public. We hope that this book can go some way towards correcting this situation.

What is presented is a set of personal views of the authors. Over the last twenty years we have seen numerous countries struggling with these conceptual and technical problems when trying to build a framework for assessing the safety of radioactive waste disposal. In a few cases, the experts involved have devised sensible, pragmatic approaches that can be readily understood by the public because they realistically take account of social attitudes and of economic feasibility. Sometimes, however, the experts have come up with logically convoluted, technocratic approaches that have led to major problems of interpretation and communication. The resulting radiation protection goals and approaches proposed for waste disposal have been impracticable and uneconomic ways to promote safety. We have tried to extract the most useful lessons from this mixed experience and to outline what we consider to be sensible ways of addressing the hard issues that face decision-makers when they have to deal with combinations of enormously large numbers (such as millions of years) and vanishingly small numbers (such as microsieverts of radiation dose). In particular, we have presented our views as a set of suggestions that might be useful to any organisation involved in setting new standards or updating old standards for radioactive waste disposal. These should also be interesting to concerned individuals in any country debating waste disposal principles and standards. Not everyone will agree with all of our suggestions, but we hope that this book will be a useful contribution to the debate.

We conclude this preface by quoting two aphorisms that we believe neatly encapsulate the tensions involved in setting standards that provide protection, not only today, but also for future generations, whilst simultaneously avoiding the inappropriate misdirection of the resources of current generations:

Out greatest responsibility is to be good ancestors ...

Today, there is often more credit given for ensuring that nothing is done wrong than there is for seeing that something is done right ...

Neil Chapman
Charles McCombie

Baden, Switzerland
June 2003

Acknowledgements

As noted above, this book has its origins in work carried out for the Japanese nuclear industry which was co-ordinated by Obayashi Corporation and carried out on behalf of, and supported by, the Japanese electric power utilities (led by the Tokyo Electric Power Company, TEPCO). We would like to thank these organisations for the resources to complete the work and for their continued interest in the study. In particular, our thanks go to Hideki Kawamura at Obayashi Corporation, and Kazumi Kitayama, formerly of TEPCO and now at NUMO, the Nuclear Waste Management Organisation of Japan.

We have been very much helped by several colleagues at the Swiss national co-operative for the disposal of radioactive waste (Nagra). Notably, Frits van Dorp, Ian McKinley and Piet Zuidema gave technical advice and suggestions for some of the text, Anne Claudel and Petra Blaser helped with the documentary research and in ensuring that the references and bibliography are comprehensive and accurate, and Urs Frick drafted the cartoons. We also express our appreciation to Sylvia Mieth of the Arius Association, who gave considerable assistance in producing the final manuscript. Any errors and inconsistencies are entirely our own. Finally, we would like to thank Pangea Resources International and the International Atomic Energy Agency for permission to reproduce text from two technical reports written by ourselves.

Contents

Chapter 1

Introduction

This review is concerned with developing principles and standards governing the safe disposal of solid radioactive wastes by burial deep in the Earth's crust, in so-called geological repositories. This management solution is advocated in the majority of countries that generate long-lived radioactive wastes from nuclear power plants or from other nuclear technologies. Although not unique in the long timescales over which their toxicity persists, radioactive wastes have focussed thinking on long-term environmental protection issues in an unprecedented way. The resources that the nuclear industry has been able to devote to examining long-term waste management issues are much greater than in other waste-producing technologies — largely because developing socially acceptable solutions for radioactive wastes has been acknowledged to be a prerequisite for continuing with nuclear power.

Despite the resources expended, there continues to be strong opposition from a significant sector of the public to the implementation of waste disposal facilities. This attitude is due amongst other things to a wide-spread fear of radiation and to the frequently insensitive response of the nuclear industry and the politician to these real concerns. This has resulted in intensive public scrutiny of radioactive waste strategies and in an especially strict regulatory framework. However, we believe that much of the debate and thinking on appropriate standards and approaches to regulating radioactive waste disposal will eventually be echoed in other environmental legislation. Consequently, the way in which principles and standards are being set at present, and the thinking behind this, are of wider interest than in the nuclear field alone. The issues are not just technical and scientific. There is also a much wider philosophical context to the debate, centring on ethics, human values and the expectations of society.

Principles precede standards hierarchically in structuring any scheme of environmental protection, and both may be enshrined in law and in regulations. Very few geological disposal repositories have yet been built, and, although the basic principles for radioactive waste disposal were formulated early (e.g. NRC, 1966;

NEA, 1982; IAEA 1983), only relatively recently has their practical application been put to the test. Today, radiation protection in general and waste disposal safety in particular are subjects on which information is freely shared internationally and which are important working areas of international organisations. Box 1 gives an overview of the more important bodies directly involved.

Standards, especially those governing radiological protection of people from practices in the nuclear industry, have been in existence for much longer (see Box 2, on the History of Radiological Protection). The last decade has seen considerable progress in the development of principles that embrace modern concepts of sustainability and in the fashioning of standards that meet the particular requirements of long-lived radioactive wastes. Radioactive wastes tend to be given special treatment in environmental protection, but the way in which principles and standards are developed and applied to them needs to be considered in the broader context of hazardous waste management.

Box 1: International Organisations Involved in Radiation Protection Matters

Several international bodies have considerable influence on developing consensus on matters of principle and their practical application in the nuclear field. Over the last 50 or more years, their work has provided the foundation for the way industry, governments and regulators approach the management of nuclear safety. Although there are many allied international groups involved in environmental protection, in the nuclear sector the most important bodies are:

International Commission on Radiological Protection: The ICRP was founded in 1928 and adopted its present name in 1950. It has a long-established link with the International Society of Radiology. The terms of reference of the Commission are to advance for the public benefit the science of radiological protection, in particular by providing recommendations on all aspects of radiation protection. ICRP is composed of a chairman and 6–12 other members, chosen on the basis of their recognised competence in the fields of medical radiology, radiation protection, health physics and radiation biology. It issued its first report in 1928. The first report in the current series, Publication 1 (1959) contained the basic recommendations approved in 1958.

United Nations Scientific Committee on the Effects of Atomic Radiation: UNSCEAR is a Committee of the United Nations General Assembly. It was established in 1955 and is composed of scientists from 21 nations. UNSCEAR has published more than a dozen reports on the levels and health effects of radiation. UNSCEAR's estimates of the health effects of radiation provide the basis for the international standards on radiation protection established by the IAEA.

International Atomic Energy Agency: Founded in 1957, the IAEA represents the interests and meets the needs of 130 Member States. It carries out its own research and provides technical cooperation in many fields of nuclear applications and is the focus of international efforts to maintain nuclear safeguards over fissile materials. The divisions of nuclear energy and nuclear safety are directly concerned with waste management safety and technology. Over the last 40 years, the IAEA has published numerous fundamental documents on safety principles and how to apply them to waste management.

OECD Nuclear Energy Agency: The NEA was formed in 1958 and is a semi-autonomous body within the Organisation for Economic Cooperation and Development (OECD). Its objective is to contribute to the development of nuclear energy through cooperation among its participating countries (currently 27 countries). It represents 85% of the world's installed nuclear capacity. It has a programme addressing issues such as nuclear safety and licensing, waste management, radiation protection, economics and technology of the nuclear fuel cycle, nuclear science, law and liability, and public information.

European Commission: The European Atomic Energy Community (Euratom) represents the interests of those countries within the European Union (EU) with nuclear power or research programmes. As part of the broader framework of EU R&D, the European Commission manages action programmes that include research into nuclear safety and waste management. The most recent of these (the 6th Framework programme) will run from 2002 to 2006. Stemming from research and discussions within the Euratom countries, the EU periodically issues guidance and requirements related to waste management and safety. The 1996 directive on basic radiation safety standards (Directive 96/29) is the most recent affecting dose limits for radiation workers and members of the public and has to be transcribed into national laws within EU member states. In 2003, a major directive on implementation of radioactive waste disposal was awaiting approval (see Appendix 1).

1.1 Wastes and Protection of the Environment

Hazardous, toxic wastes have been generated on a large scale since the start of the industrial revolution, more than two hundred years ago. For a long time, little

Box 2: The Development of Radiation Protection

The existence of radiation has been known for only a little over a hundred years, since Röntgen discovered his "X-rays" in 1895 and, shortly afterwards, Becquerel discovered radioactivity, Marie Curie discovered polonium and Pierre and Marie Curie discovered radium. Although harmful effects were identified and associated directly with X-rays almost immediately, it was some decades before the hazards of naturally radioactive substances were widely recognised (Lindell, 1996).

It took many years to develop the concept of radiation doses to people. The early decades of the 20th century saw interest focussed (through the International Congress on Radiology) mainly on developing methods to measure radiation, with the Röntgen unit (r) of incident radiation being established by the International Commission on Radiation Units and Measurements (ICRU) in 1928. The International Commission on Radiological Protection (ICRP) was formed in the same year. The concept of "tolerance dose" was developed and, in 1934, ICRP made its first recommendation of a tolerance dose of 0.2 r per day: about 500 mSv/a. This can be compared with the currently recommended dose limits of 1 mSv/a for members of the public, or 20 mSv/a for occupational doses. Tolerance dose was the first application of the principle of dose limitation.

With the development programme that led to nuclear weapons in the 1940s came substantial increases in understanding of radiation effects on people. Health physics, as a distinct branch of medicine, originated within the Manhattan Project. Within the USA, an advisory committee recommended a tolerance dose of 0.1 r per day, and the National Bureau of Standards proposed that body contents of more than 0.1 microgrammes of radium were unsafe. Until only a few years previously, radium preparations had been widely advertised commercially as having therapeutic effects. The implications of a "tolerance dose" were challenged, and the idea of a "maximum permissible dose" (MPD) proposed instead.

In the years following the Second World War, maximum permissible concentrations (MPCs) of radioactive substances in air and water were derived for the first time (1953), based on MPDs. The same year saw the first UK–USA–Canada agreement on a dose limit specifically for members of the public, at 1.5 r per year. Sievert in Sweden and Spiers in the UK began work to quantify natural radioactivity as a basis for deciding what would be suitable permissible incremental doses for people, above natural exposures. ICRP was reconstituted after the war, issuing its first real publication in 1955 (although not in the famous "Publication" series) which began to look at permissible doses to various human organs.

The concept of MPDs for members of the public came from the ICRP in 1956. The limit was in the range of variability of doses from natural, background radiation. It was a fraction (10%) of the dose limit recommended for radiation workers. The same year saw the first work of UNSCEAR (United Nations Scientific Committee on the Effects of Atomic Radiation), which produced its first report in 1958. The concept of "genetically significant doses" was introduced, which was a first recognition that radiation could have stochastic effects in the population as well as deterministic effects to individuals. Since the inception of this important concept, stochastic effects have been assumed to have no lower dose threshold, an issue that still causes considerable argument. The IAEA was founded in 1957, with its nuclear safety division centrally concerned with the impacts of nuclear power and weapons programmes.

ICRP's Publication 1 (1958) recommended an annual MPD for members of the public of 0.5 rem (5 mSv). It reiterated that the most conservative approach was to assume no threshold and no recovery, so that even low accumulated doses would induce leukaemia in some susceptible individuals. They also emphasised that MPDs should be regarded as maximum, and that all doses should be kept as low as practicable. By 1965, the latter words read "...as low as is readily achievable, economic and social consequences being taken into account" (ALARA in ICRP Publication 9). By this time, the emphasis of radiological protection was firmly on the stochastic effects of radiation and the issue of "risk" came to the fore, where it was formalised as "synonymous with the probability of death". First attempts at defining cost–benefit analysis, or how much money it would be reasonable to pay to eliminate a unit radiation dose, were made in 1970.

In 1977 the unit *sievert* replaced the rem (1 rem = 10 mSv) in ICRP Publication 26, which also introduced the three basic rules of justification of a practice, optimisation of protection and individual dose limitation. ICRP 26 was used as the basis for the Basic Safety Standards issued by the IAEA and other organisations. The first ICRP publication specifically to deal with radioactive waste disposal came in 1986 (ICRP 46). The major milestone Publication 60, which superseded ICRP 26, forms the current foundation for radiological protection, and is the starting point for this current review. At the time of writing ICRP is entering a new stage of discussions with a view to updating ICRP 26 in about 2005.

The reader may well be struck by the progressive reduction in dose limits that are recorded in the above description: by a factor of about 100 between 1934 and 1958, and a further factor of five since then. In the first period, this was caused by better knowledge about the deterministic effects of radiation. Since 1958, the reduction has been due to increased knowledge about the stochastic effects of radiation, derived from those exposed to the Japanese atomic bombs, and from other sources. The reduction also reflects the considerable conservatism or cautiousness of the radiological protection community.

thought was given to their disposal; the world was infinite, the "dilute and disperse" approach in which the objective is to reduce concentrations of pollutants by dilution in larger quantities of air or water seemed suitable. Only in the last forty years has the awareness grown that human activities can indeed damage the global environment, with the publication in 1962 of the book *Silent Spring* by Rachel Carson providing a powerful warning message. Even today, only the most-developed countries seriously attempt to ensure that wastes are managed so as to minimise the potential for environmental pollution. In many, less-developed countries, there are still only limited resources available to reduce the present and future health and environmental impacts of waste disposal. As a consequence, all industrialised countries possess legacies of current or historic poor waste management practices that provide concerns for the future.

With the exception of some gaseous discharges, all wastes that are released to the environment find their way either into the ground or into surface waters. The main historic polluters include the mineral extraction industries, building and construction, metal smelting and refining, coal and gas production, the chemical and hydrocarbons industries and numerous small and specialised manufacturing industries handling toxic materials. There is also a growing quantity of domestic household wastes in most countries, some of which is either toxic in nature or which can produce toxic substances as it degrades. Measured by volume, the long-lived radioactive wastes that are the subject of this book represent only a tiny fraction of the wastes that need to be managed (Fig. 1.1).

In the developed countries, efforts to control environmental impacts of wastes have centred on legislation to limit atmospheric and liquid discharges and to ensure that solid wastes are routed to appropriate landfill or other burial sites, and that such sites are properly managed during and after operations. In the jargon of the waste community, there has been a movement towards the "concentrate and contain" strategy for waste management. The development of environmental protection over the last few decades has involved the progressive introduction of quantitative standards, the majority of which tend to be based on the simple principles of protecting the health of current generations of people against the direct toxic hazard presented by wastes. Only recently have standards begun to consider other less obvious risks, such as stochastic risks (where the probability of harm rather than the severity is governed by exposure), direct risks to future generations, and risks of genetic consequences to future generations. Many countries have environmental standards for the protection of air quality and of water supplies, either from discharges of liquid effluents or from leaks from solid

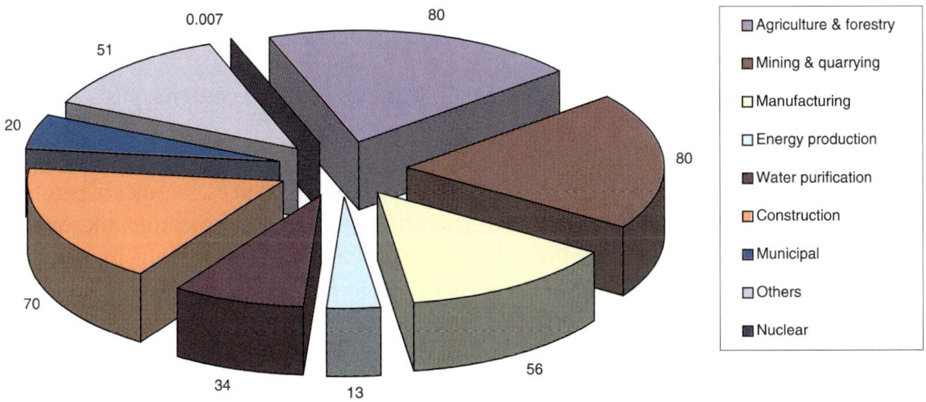

Fig. 1.1. Reliable information on worldwide waste arisings is not available. As an indication of the relative significance of radioactive wastes compared with other wastes, this diagram shows the annual production of all types of waste by source, in a typical developed country with a significant nuclear power programme (the UK). The values show millions of tonnes of waste. It can be seen that the mass of radioactive waste (7000 t/a) is a minute fraction, even of the energy production wastes (which are mainly fly-ash from coal burning). Less than a quarter of this small amount is long-lived wastes. It is recognised, of course, that the mass of wastes produced is itself not a sufficient criterion for judging the magnitude of the disposal problem. (Information from OECD, 1997.)

waste disposal sites. These are usually in the form of emission standards (acceptable levels of release), quality standards (acceptable concentrations in air, water or soil) or exposure standards (acceptable exposures or doses to people). We discuss chemical risks and standards in relation to radiation risks and standards in Chapter 13.

Understanding of health impacts on people exposed to chemo-toxic materials (e.g. dose–response relationships) is still rudimentary for many substances, as is knowledge about the behaviour and fate of some pollutants as they move through soil–water systems and the biosphere. Long-term ecotoxicity effects in the natural environment, in terms of impacts on single species and whole ecosystems, are only now coming to be better understood. Waste continues to be produced in increasing quantities and diversity, and regulation, and the scientific understanding on which it must be based, are still catching up. In the 1970s and 80s, the situation worldwide was one of fire-fighting; running to catch up with the impacts of past practices. Tragic cases, such as the Love Canal in Niagara Falls, USA[1] illustrated dramatically that improper disposal of toxic wastes can lead to human suffering and expensive

[1]Almost a thousand families had to be evacuated from a development area sited on top of a 30-year-old landfill containing ~20,000 tonnes of chemical wastes, when contaminants began to reach the surface following a rise in the water table.

remediation requirements. The situation is still imperfect today for some environmental contaminants, particularly when one takes a longer-term perspective and tries to identify and remedy problems that may become apparent or acute only in the distant future.

Recently, governments and international organisations have begun to look at broader principles that could be applied to industrial practices, so as to provide protection both now and in the future. Focusing purely on protecting the health of the current human population has been recognised to be an inadequate response. Environmental impacts across many species may be important, and the fact that many contaminants can be persistent in the environment means that a much longer timeframe needs to be taken into account. Growing awareness of the need to protect future generations has driven much of the work on the development of safe disposal methods for the radioactive wastes that are the subject of this book.

More emphasis is placed on another approach to waste management, namely "reducing and recycling". Waste reduction requires cleaner processes that do not produce as much, or as hazardous, waste. The overall objective is to minimise the amount of waste that will be produced by a practice; what emerges at the end is a minimum amount of material with no realistic commercial value. Lack of value might be simply as a result of intrinsic lack of worth, or because the material is too contaminated or difficult to recycle. The order of consideration, when either developing a new industrial process or practice, or back-fitting environmental controls to an old one, is:

- Select a process or system design that produces little or no waste.
- If waste must be produced, design the process to be as efficient as possible and minimise the amount of waste produced.
- Recycle as much waste material as possible.
- Treat and condition any remaining wastes to reduce volumes and to put them in a largely inert, stable physicochemical state for disposal.
- Select a final disposal route that will not cause long-term environmental problems.
- Demonstrate as openly and convincingly as possible to the experts, the regulators, the politicians and the public that the required environmental protection levels will be achieved.

An approach of this sort fits well with the *sustainability principle* — the currently much-discussed goal of passing on to future generations as many as possible of the freedoms we enjoy and as few as possible of the problems that we can solve ourselves. As well as for management of current practices, similar principles have been developed for the introduction of new technologies, where the *precautionary principle* (not embarking on a potentially hazardous path unless the hazards are well understood: see Chapter 3) is intended to provide proper safeguards against causing uncontrolled harm — without, however, simply blocking all technological progress.

In this context, there is, for example, currently a heated debate surrounding the introduction of genetically modified organisms into the environment.

Principles like those mentioned above are based on ethics, applied to the way in which we treat our environment. They provide a top-level starting point for building a framework of subsidiary principles and, eventually, standards for environmental protection from the impacts of wastes. Whilst standards (e.g. for drinking water quality) exist in many countries, the development of a unified framework is still in its infancy, particularly with respect to integrating national policies to deal equitably with the risks from all types of environmental hazards. Standards exist for some aspects of pollution control, regulations exist to govern certain practices, but the basis is not yet comprehensive or consistent.

1.2 Radioactive Wastes

In the midst of these general environmental developments, industries based on the application of nuclear technologies and those organisations concerned with radiological protection have been addressing the same types of issues, with a strong emphasis on waste reduction and environmental protection. In fact, radiation protection in waste disposal has been in the vanguard in terms of the rigour of standards and of the associated regulations. To some extent, the intensive work on radiation protection has been a reaction to the well-recognised phenomenon of "nuclear phobia". Radiation, often characterised as being incapable of being touched, seen or smelled, has always aroused public fear and apprehension (Weart, 1988).

HE'S GROWN A FOOT SINCE I SAW HIM LAST....

Nevertheless, in radioactive waste disposal too, there are still areas in which there is no common view of the most appropriate approaches to very long-term protection of people and the environment, or of the appropriate standards to use.

A considerable literature has developed as aspects of radiological protection and nuclear industry practices have come under scrutiny, or as nuclear processes or facilities have needed to be licensed. A key area is that of disposal of long-lived radioactive wastes, not because this area presents the biggest hazards, but because it highlights the controversial long-term issues. Long-lived radioactive wastes include:

- spent fuel (SF) from nuclear power reactors;
- wastes from reprocessing spent fuel to extract re-useable elements, in particular the high-level wastes (HLW);
- some wastes from operating nuclear facilities;
- materials from production and dismantling of nuclear weapons;
- some of the construction materials from decommissioned reactors and nuclear plants;
- some of the radiation sources and other radioactive materials used in medicine, research and industry.

Not listed here, and not treated in detail in this book, are the long-lived tailings from mining and milling ores of natural materials, such as uranium. The large volumes of such materials and the fact that they are naturally occurring leads to different approaches being employed for their management. From the outset, when substantial amounts of wastes began to build up soon after the start of commercial power generation and during the intensive nuclear weapon production programmes of the Cold War, it was realised that special provisions would need to be made for their disposal. Although there were numerous early cases of (generally) small discharges of long-lived wastes in western countries, often in liquid form, storage in conditioned, solidified form soon became common practice. This was only possible and economic because the absolute amounts, compared to wastes from other energy-generating processes, were very small. Even today there are less than a million cubic metres of long-lived wastes in storage worldwide, although this will begin to rise sharply as older nuclear facilities begin to be decommissioned[2]. This waste has been produced by over 400 nuclear power plants that are operating around the world, with a total capacity of 358 Gwe, which supply about 17% of the world's electricity. It is informative to compare this with the wastes that would be

[2]Obtaining an accurate figure for global long-lived waste quantities is made difficult by differences in classification and estimation. Information from the IAEA Waste Management Database, combined with other published sources (Ahearne, 1997; UK Nirex, 1999) allows one to estimate a figure of between half a million and a million cubic metres of long-lived intermediate level waste (LL-ILW) in store worldwide and what would amount to some tens of thousands of cubic metres of HLW and SF when conditioned for disposal. Although most of the volume is in ILW, the bulk of the activity is in the much smaller volumes of HLW and SF.

produced *each year* by coal-burning power stations with the same capacity: over one hundred million tonnes of ash (including ~140,000 tonnes of heavy metals) and over two thousand million tonnes of carbon dioxide.

Long-lived radioactive wastes will remain significantly radioactive for immense periods of time: of the order of hundreds of thousands of years or more. The longest-lived radionuclides in the wastes will remain active for millions of years. The long — but finite — hazardous lifetimes of radioactive wastes are viewed as a huge problem by some commentators, who often overlook the facts that the specific radioactivity of a nuclide with a very long half-life is low and that other toxic wastes, such as heavy metal residues, will be toxic forever. As a result, radioactive waste management is the first case where the need to consider such very long times has come so forcibly to the attention of those concerned with providing environmental protection. More than 40 years ago, geological disposal was proposed as the technical solution for providing proper protection for the environment (NAS, 1957). A diversity of alternative options have been examined (e.g. BNWL, 1974), but today geological disposal is still recognised to be the only viable approach that avoids the burden of ensuring safe and secure perpetual storage (NEA, 1999a; NRC, 2001a). Decades of effort have gone into developing suitable geological disposal solutions, with the focus being always on evaluating the long-term safety of radioactive wastes emplaced in deep underground repositories.

Much of the emphasis on managing long-lived wastes has been focussed on ensuring that they can be disposed of safely, without consequences for the environment. More recently, another issue has arisen, concerning security: "safe-guarding" the management and disposal of wastes that contain substantial amounts of fissile radionuclides, which could be used for making nuclear weapons. The terrorist attacks on the USA in September 2001 added a further factor, causing considerable reflection on the security of all types of surface nuclear facilities. Taken together, these safety and security considerations reinforce the logic of moving wastes out of vulnerable surface stores into secure underground repositories in suitable geological formations.

In essence, geological disposal aims to isolate the wastes in engineered barriers in deep, stable rocks, so that almost all the radioactive material they contain will decay within the repository or immediately surrounding rock. This will take a few hundred years for some radionuclides, millions of years for others. At no future time should radioactive nuclides return to the human environment in concentrations that could cause unacceptable hazards, by today's standards. However, over these timescales it is scientifically impossible to rule out that some radioactivity, albeit in minute amounts and concentrations, will migrate into regions of the rock, groundwaters and biosphere where people may come into contact with it, far into the future. Also the "concentrate & contain" approach inevitably results in there being local concentrations of the longer-lived radionuclides, which will remain for millions of years and be analogous to a uranium ore body.

Safety analysts for geological repositories spend almost all of their time and effort trying to quantify as well as possible the tiny fractions of the radioactive inventory which will ***not*** be completely contained in the deep underground until they decay away to total insignificance. This distracts attention from the fact that the passive, stable safety barriers will contain the waste for so long that most radionuclides will decay in place. Disposal in deep geological formations was proposed precisely because the geological environment at these depths can remain effectively stable and unchanged, even over the extremely long radioactive lifetimes of the wastes. The essence of safe disposal is to be assured, with reasonable confidence, that no component of the waste will pose an unacceptable hazard to people or the environment at any time. Providing this assurance is not an easy task. The convolution of long times, complex natural environments, debatable radiological effects of low levels of radiation and the social question of what we should reasonably expect of a safe system are the subjects of this book.

The current consensus is that the proper approach is to try to find solutions that will give complete protection for as far into the future as we reasonably can, and to avoid practices that would expose people at any future time to hazards that we are not prepared to accept today. This is not a self-evident conclusion, however. It is reasonable to question how much we should worry about what happens in ten thousand years time or whether we should devote scarce, present-day resources in trying to protect hypothetical people hundreds of generations in the future from very small risks, while current societies face bigger, more immediate problems (see Chapter 2).

Many countries and organisations are thus struggling at present with:

- defining what comprises adequate protection now and in the future;
- developing deep geological disposal solutions for long-lived wastes that can offer this degree of protection;
- carrying out evaluations of the future behaviour of disposal systems that can be accepted with sufficient confidence by decision-makers.

While there is, as mentioned earlier, an extensive literature to guide such groups, and a great deal of consensus on many issues, both qualitative and quantitative, there are still numerous aspects requiring further debate and interpretation. A key topic concerns the different ways of framing objectives and regulations, since these can have large impacts on the way in which disposal is carried out.

This book endeavours to bring together all the larger issues concerning guiding principles for safe disposal of long-lived wastes. Consideration of appropriate ways of addressing these issues is crucial for any country or organisation starting afresh with setting up principles. In practice, few have the luxury of starting from scratch, as they carry with them the legacy of decades of policy and regulation, developed for specific purposes. The topics addressed, however, should also be of direct interest for those reviewing existing systems or concerned with making these systems transparent to a wider public. Apart from reviewing the main areas of debate, we have thus concluded by abstracting from them a suggested "state of the art" set of factors and positions that we believe could form components of a sound basis for any set of national regulations.

The debate on safety principles will continue over coming years, especially in radiological protection circles. This is assured by the unabated public interest and debate on all things nuclear. Thus, we would fully expect the present conclusions to need revision in the future. However, we believe that it is unlikely that future developments will give rise to major changes in approach or objectives of disposal and its regulation. What we would hope, is that the issues that we believe have been successfully resolved for radioactive waste management might be translated to the management of other wastes and other environmental hazards, resulting in a more equitable approach to deploying the world's resources to protect future generations.

1.3 The Need for and Structure of Safety Principles

Is it necessary, or even worthwhile, to try to establish a wide public and political consensus on overriding principles governing radioactive waste management? Without well-defined principles, we believe, it is difficult or impossible to develop rational and defensible policies for waste management, or sensible standards against which to gauge practices. In such a situation, one constant temptation for decision-makers will be to avoid decisions, to postpone new developments and continue with unsatisfactory interim practices. Agreed, transparent principles are important prerequisites for involving wider circles of stakeholders, including the public, in societal decisions on how to offer adequate environmental protection. Alternatively, lack of a proper framework can lead to demands for unnecessarily expensive solutions to poorly perceived, trivial or non-existent problems. Principles provide fundamental guidance by encapsulating the basis, both ethical and technical, of why things need to be done; standards provide yardsticks so that we can tell whether they have been done properly.

Principles are thus ethical and technical (or scientific), with the ethical principles providing the lynchpin for the rest. In waste management applications, there should be a hierarchy of principles and standards, moving from what should be easily understood statements of intent that must attract wide endorsement, down to much more detailed technical standards that cover what is expected in practice of a waste disposer. As noted above, principles precede standards, although, in an ideal situation, these principles and standards would lie within a common framework of environmental protection for a nation or a community. Consequently, a simple hierarchy can be defined:

- **ethical principles** (e.g. sustainability, protection of the environment and of people, protection in the future as well as today, equitable use of resources, etc);
- **technical principles** (e.g. use of a systematic approach covering all aspects, time periods to be considered, types of waste concerned, hazards to be protected against, etc);
- **standards** (measures of acceptability of a practice or a proposal: radiological and other)

- **regulations** (incorporating both principles and standards and explaining how they are to be applied);
- **guidelines** (giving advice on how standards and regulations can be complied with).

The next section looks in more detail at the concept of geological disposal of long-lived radioactive wastes and addresses some of the problems facing those developing standards. After this background discussion, we consider in depth the following issues, which all need to be considered when defining principles or setting standards and moving towards a set of regulations:

- **Ethics:** what is our responsibility towards present and future generations in terms of providing health and environment protection while making effective use of scarce resources?
- **Retrievability:** do we understand repository systems well enough to fill them with waste and seal them in the safest and least recoverable way, or should we allow for the whole process to be reversible at any time?
- **Timescales:** how far into the future should we provide protection and how do we deal with times outside human experience?
- **Performance measures:** which parameters should we use to assess whether a repository is safe and how might these be incorporated into standards?
- **Siting a repository:** which standards and criteria should guide the process of choosing a location for a repository?
- **Disruptive events:** how should we consider, give weight to and put into social context the possibility that major natural events might affect the behaviour of a perfectly constructed repository?
- **Human impacts:** to what extent should we consider the possibility that people may interfere with a repository in the future, when we are estimating long-term safety?
- **Monitoring and control:** how should we actively control a repository site after it has been closed: what do we need to monitor during and after the operational life of a repository and will this help to assure long-term safety?
- **Preserving records:** how should a repository site be marked and records of its nature be maintained so that knowledge is not lost after active controls have ceased?
- **Uncertainty:** given all the obvious uncertainties surrounding these issues, how can standards adequately take them into account so that decisions based on uncertain numbers have some meaning?
- **Chemotoxicity:** are there useful parallels between the way we manage and regulate chemically hazardous substances and radiotoxic substances?

1.4 Responsibilities for Setting Principles and Standards

Internationally, it is recognised that countries with radioactive waste management responsibilities (in particular with wastes from nuclear power production) must have a proper institutional framework allocating responsibilities. Principle 6 of the IAEA Safety Fundamentals (IAEA, 1995a) is that radioactive waste shall be managed within an appropriate national legal framework including clear allocation of responsibilities and provision for independent regulatory functions. National legal requirements may need to reflect international agreements and conventions (see Appendix 1). Legal and governmental responsibilities within the radiation and nuclear safety sectors are not, of course, restricted to waste management. The IAEA notes (IAEA, 1995b) that many responsibilities are common to a broad range of facilities and activities, including research, industrial and medical uses of ionising radiation, mining, processing and transport of radioactive materials, nuclear fuel manufacture, nuclear power and research reactors, industrial irradiation facilities, and the decommissioning of plant and rehabilitation of sites. Individual organisations within a national infrastructure might be responsible for safety aspects in several or all of these areas.

So far as the waste management area is concerned, a national framework would consist of a **national policy** for waste management and environmental protection, a **strategy** for implementing this policy and a **waste management system** including procedures and facilities for storing and disposing of wastes, for regulating activities and for monitoring environmental impacts. The components in each category are described briefly below.

- **National Policy** should define, in legal terms, the broad environmental objectives required of radioactive waste management, specify whether disposal is required and the extent to which an integrated environmental approach is to be developed, legislate the powers of regulatory agencies map out and the nature of public and other stakeholder interaction desired in decision-making processes.
- **National Strategy** should develop broad timescales for implementing disposal, specify a funding mechanism to ensure that disposal, regulation and all necessary planning, consultation and decision-making procedures can be carried out properly, define the responsibilities of each private and governmental organisation concerned and the procedures to ensure that necessary interactions take place. In some countries, the national strategy may be closely linked to national policy and both encapsulated in law.
- **Waste Management System** should comprise the organisations that would implement the strategy and the facilities that they will need to do this: e.g. for waste treatment and storage, technical and scientific R&D, repository construction and engineering, environmental characterisation and monitoring.

Within this framework, the responsibilities for waste management are divided between the **State** (government), the **regulator** and the **waste producer** (who, in several countries, passes this on to the specific body charged with managing radioactive wastes: the operator or **implementor**). The duties of each are as follows:

- **State (government) responsibilities:** establish the legal policy framework, the strategy, and appropriate regulatory agencies, and ensure that adequate resources and infrastructure are available to manage the wastes, carry out any necessary R&D and give the necessary continuity of existence to the waste management system.

- **Regulator responsibilities**: define safety principles, criteria and regulations, enforce compliance by the implementor and waste producer with legal requirements, implement licensing processes and provide authorisations for waste disposal facilities, and advise the government. The regulator may also have responsibility for long-term monitoring, independent testing and inspection, and enforcement of corrective actions.
- **Implementor (& waste producer) responsibilities:** manage waste safely at all stages (the implementor has the primary responsibility for safety), identify appropriate disposal solutions, select suitable disposal sites and repository systems, demonstrate their technical adequacy and implement disposal (and subsequent site management) to the satisfaction of legal and regulatory requirements.

As noted above, when establishing the legal framework and the regulatory body that will execute the legal actions, the State should, in principle, aim at a system that is compatible with other regulated activities. There might thus be some overlap of responsibilities (e.g. for the routine health and safety of operators). In practice, there is often a tendency at the political level to place special requirements on radioactive waste management; requirements that reflect more the widespread aversion to nuclear matters and less the objective hazards associated with radioactivity. Those responsible for implementing waste disposal should be partners with the regulators in preparatory discussions and they should feel obliged to insist on even-handed treatment of nuclear issues.

All bodies involved in setting principles and standards need to operate in a transparent and open fashion. In the UK, the Royal Commission on Environmental Pollution notes (RCEP, 1998) that this involves giving full publicity to the existence of the bodies, their terms of reference, the decisions that they take and the reasons for them. There must be adequate opportunity for all of those outside a body, but with an interest in a given decision, to contribute fully to the decision-making process.

The RCEP also notes that bodies setting standards should:

- have their decisions informed by an understanding of people's values, which should be articulated as early as possible when setting standards and developing policies: the public should be involved in the formulation of strategies rather than merely being consulted on already drafted proposals;
- draw explicit distinctions between scientific statements and recommendations that they wish to make after considering a scientific assessment in conjunction with other factors;
- establish an audit trail documenting all considerations taken into account in reaching a decision;
- review standards at pre-set intervals or earlier if significant new evidence emerges or there is an unforeseen change in circumstances;

- have all their analyses subject to peer review;
- be able to relate their decisions to decisions about other environmental risks within the geographical area they cover.

The issue of policy setting and the process of putting in place and carrying out a repository development programme for the disposal of long-lived wastes is discussed at several points in this review, and returned to in the conclusions.

1.5 Stakeholders and their Role in Setting Principles and Standards

Who are stakeholders? The term is widely used in all decision processes today, including those involving waste management issues. Originally, the term referred to a neutral person holding the stakes or wagers of people betting until the result of the game or competition was known, and then distributing the winnings. A second meaning, more relevant to its current application, refers to people who have a share or interest in any enterprise, this interest usually being financial.

A recent Euratom study (European Commission, 2000) defines stakeholders as individuals or groups who are promoting their own interests and are not responsible for reaching a balanced solution. The Commission separates from these stakeholders the public authorities whose views, it is claimed, must "converge to defend the interest of society, not their own interests". The USDOE similarly distinguishes itself and the regulators, USNRC and USEPA, from stakeholders who are members of the public (English, 2000). The US National Research Council (NRC, 1996) avoided the term stakeholder and used instead "interested and affected parties". They defined "affected parties" as people, groups or organisations that may experience benefit or harm as a result of a hazard, and "interested parties" as people, groups or organisations that decide to become informed and involved (but who need not be affected parties). The NRC differentiated between these interested and affected parties on the one hand and the decision makers and "neutral" risk analysis specialists on the other.

These idealistic and rather elitist views of the scientific bureaucracy and the establishment are contradicted by Webster (2000) summarising the views of attendees at a NEA arranged Forum on Stakeholder Conference (NEA, 2000a). There the consensus was that the regulators as well as local and national officials should also be regarded as stakeholders who will have their own biases and interests. In the same Forum, the spatial and temporal dimensions of the stakeholder issue are emphasised (English, 2000). This distinction is particularly relevant in waste disposal. Here, community, regional and national interest often give rise to a "doughnut" effect, in which acceptance of a repository is often lowest in the intermediate region where local benefits are not apparent and the national interest is given less weight. Furthermore, in waste disposal the temporal dimension leads to problematic discussions on who speaks for the generations of future stakeholders who can have no current voice in the debate on siting.

In practice, there has been a rapidly growing acknowledgement that active participation of a wide range of stakeholders is one way to work towards democratically responsive environmental regulation. In many case studies, specific processes for identifying relevant stakeholders have been introduced; in the most extreme cases anyone who chooses to become involved automatically becomes a stakeholder. Once public stakeholders are identified, there is a whole range of procedures for organising their input. These include public hearings, citizen advisory committees and task forces, alternative dispute resolution techniques, citizen juries or panels, opinion surveys, focus groups, etc. (NRC, 1996).

Interesting studies have been done on the effectiveness of this input — and in particular on the use of science within stakeholder processes (see, for example, EPA, 2001a). In general, the conclusions are that stakeholder processes can be valuable in supporting science-based decision-making of the type involved in regulating radioactive waste management. Sufficient time and effort must be devoted to the stakeholder process, however, and sufficient commitment to the process must be shown if these positive results are to be achieved.

There are also some major caveats to the conclusions:

- Stakeholder processes should not be relied on uncritically, given the limited resources and capabilities of some of the groups involved;
- Stakeholder processes should be used only with caution and with appropriate safeguards in the regulatory area, if practicable and equitable safety standards are to be attained;
- Scientists involved in risk-informed decision-making should not abdicate their responsibilities for ensuring that only good science is used and that no relevant good science is ignored in the decision process.

This concluding point does not ignore the fact that scientists are often not neutral and non-affected parties in the debate. In particular, when the scientific evidence itself becomes a matter of controversy, then those scientists involved in assembling this evidence obviously becomes potentially affected by the loss of credibility if their science is found lacking.

1.5.1 Stakeholder Interactions: Current Status

The state, regulatory and implementing bodies are three very important stakeholders in the general area of radioactive waste disposal, and in the particular process of setting standards. They are not, however, the only stakeholders. One of the most significant developments in waste disposal planning over the last years has been the sharp increase in efforts to ensure that all interested and concerned parties have a right not only to be fully informed but also to be involved in the actual decision process. This development was overdue. The nuclear industry in general grew out of a background based on maintaining secrecy, whether military or commercial. Decisions on technologies, organisations and facility siting were taken in closed circles and then made known ("decide, announce, defend"). Politicians did little to influence this and more or less subscribed to technocratic approaches so long as public opinion was not aroused. Today, society seemingly wishes to be more directly involved in all contentious issues and those responsible for waste disposal planning have realised this.

One can make a convincing case that a large share of the responsibility for the progress towards implementing geological disposal being so much slower than was originally imagined is borne by those who failed to appreciate the importance of stakeholder interactions and the political paralysis that has resulted from the widespread inability of governments to embrace these. It is widely acknowledged that the problems facing geological disposal are more societal than technical (NEA, 1999b; NRC, 2001a). Individual national disposal programmes such as those in Canada, the UK, Spain and Germany have been almost completely stopped, owing to the inability of the implementors to win sufficient political and public support. In other countries (e.g. USA, France, Switzerland and Sweden) specific projects have suffered significant delays for the same reason. Belatedly, the problem has been recognised and efforts to improve matters are underway in many countries (e.g. DEFRA, 2001). The international organisations have also awoken to the fact that science and technology are not enough and the NEA, the IAEA and the EU are all currently developing concepts for improving stakeholder confidence in waste disposal (e.g. NEA, 2000a, 2001a).

There is still a long way to go, however. Even official non-governmental organisations (NGOs) have difficulties gaining entry to working groups and meeting of the international bodies. For private individuals, it is often impossible. At the national level, opportunities for participation are sometimes greater. In some countries, the legal procedures governing promulgation of new laws and regulations guarantee individuals the right to be informed and to comment. The USA has a formal process of rulemaking that encourages public input, even to relatively detailed aspects of proposed regulations. Waste management organisations are now devoting more of their time and energy to creating and maintaining links to the wide variety of stakeholders.

It is important that this trend continues and that the responsible bodies actively encourage encompassing the stakeholder circle. A reluctant or grudging opening of the doors only to those who are outside, pushing hardest, could be counter-productive. For example, whereas various NGOs have an obvious right to be part of the debate, their mission is openly acknowledged as being advocates of a particular partisan view. They certainly cannot claim to represent the public in a better way than is done through the normal democratic representation process. The appointed officials in this process can legitimately claim to represent the public. In a controversial, polarised issue such as waste disposal, however, they have a continuing responsibility to seek wide public input.

Although we certainly regard wide consultation and the opportunity for participation in setting standards to be essential, we conclude this chapter on a note of caution. The single act of enlarging the field of involved organisations will not ensure successful outcomes endorsed by all. Hard work and innovative

A common view of participation...

thinking are needed, as well as continuing enlightened leadership that does not abrogate the political responsibility for making difficult decisions. Simply turning difficult issues over to the public is not enough. The following quote comes from Kasperson (2000):

> ...Currently, we are on the stakeholder-involvement express, barrelling down the rails of well-intentioned but often naive efforts to address growing public concerns over risks, changed public expectations over the functioning of democratic institutions, and historic declines in social trust in those responsible for protecting public safety.
> ...We know relatively little about which participatory interventions are likely to be successful, or even what success means, in different communities and social settings.
> ...perhaps it is time to put the brakes on the current stakeholder express, or to switch to the local, so that these processes become much more reflective and self-critical, that they are goal- not technique-driven, that they are rigorously evaluated by independent parties, that potential abuses (e.g., kicking controversial issues to public) are controlled, and that they are accountable to and collaborative with those in whose name the experiments are mounted.

In summary, the task of structuring stakeholder involvement depends on mastering a delicate balance between having an inclusive process involving both experts and non-experts and avoiding the abdication of their responsibilities by elected officials or appointed experts.

Chapter 2

Safety and Security Issues in Deep Geological Disposal

The aim of deep geological disposal is to isolate radioactive waste from the biosphere[3] and the everyday activities of people until it presents no significant hazard. By this, we mean that any negative impacts will be no greater than those that we are prepared to accept in the natural environment. Geological disposal works by ensuring that as much radioactivity as possible decays without ever reaching the biosphere and that any radioactive nuclides that could do so would be so dilute (with releases dispersed over long periods in the far distant future) as to be of no radiological concern for people or the environment.

Protecting people involves more than isolating the radioactivity of the wastes. As noted in Chapter 1, there is also a security aspect associated with isolating wastes that contain fissile radionuclides, which could be used in nuclear weapons. If disposed of, such materials need to be made irrecoverable by any practicable clandestine means.

Very few deep geological repositories exist at present. Countries with substantial nuclear power generation programmes are carrying out research into siting and constructing such facilities, but there are widespread problems of gaining

[3]International radioactive waste management safety studies have developed their own, rather specific terminology over the last 20 years. Deep geological repository projects generally refer to three simplified regions of the "repository system":

- the *repository*: the underground openings for access and waste emplacement and the engineered barrier system (EBS), comprising the waste itself and the components placed around the waste;
- the *geosphere*: the rock formations in which the repository is constructed, and those surrounding and overlying the host rock formation up to the ground surface;
- the *biosphere*: the near-surface and surface natural environment in which people carry out their everyday activities.

societal (and, consequently, political) acceptance that disposal should go ahead. The concept has been evaluated exhaustively over the last 25 years, and the earliest considerations of geological disposal go back as far as the 1950s (NAS, 1957).

This book is devoted to detailed consideration of key issues affecting how deep disposal can be properly carried out and, in particular, properly regulated. To give a common background to these considerations, in this chapter we give an overview of the concept of how waste is to be contained and consider how environmental impacts of disposal might occur, and how security issues might arise. Estimation of environmental impacts produces qualitative and quantitative information that is used to assess whether a proposed repository would be acceptably safe. Acceptable safety and the wider question of acceptability in general, must be judged against the principles and standards with which we are concerned here.

2.1 What are Long-lived Radioactive Wastes?

Geological disposal is the preferred option advanced for categories of waste with high levels of radioactivity and/or significant contents of long-lived radionuclides. The categorisation of long-lived wastes is as follows:

- spent fuel (SF) from nuclear reactors;
- high-level waste (HLW) residues from reprocessing spent fuel;
- long-lived intermediate-level waste (LL-ILW) from various sources.

Fresh nuclear fuel has low levels of natural radioactivity before it is used in a reactor, and can be handled without the need for radiation shielding. Most reactor fuel is in the form of fuel elements comprising pellets of ceramic uranium dioxide sealed within thin metal tubes (e.g. of stainless steel or zircalloy). A number of these elements may be bundled together in a fuel assembly. One or more spent fuel assemblies would be sealed into a metal container for emplacement in a repository. Figure 2.1 illustrates one example of the numerous existing designs of fuel assembly and a disposal container.

After it has been involved in the nuclear fission process, the fuel becomes intensely radioactive, largely as a result of the formation of other new radionuclides: "fission products" — see below. With time, the build-up of fission products within the fuel reduces its efficiency and, after a few years, it must be removed from the reactor (becoming "spent fuel") and replaced. At this time, the original enrichment of fissile uranium 235 in the fuel (3–5%) has been reduced to about 0.8% and the content of fission products and newly formed heavy elements, including plutonium isotopes, is about 5%.

HLW originates as a liquid residue from reprocessing SF to extract the uranium and plutonium for reuse. The liquid contains most of the radioactivity from the original SF. It is commonly evaporated to dryness and the residue containing the radionuclides is then melted with a much larger volume of inert borosilicate

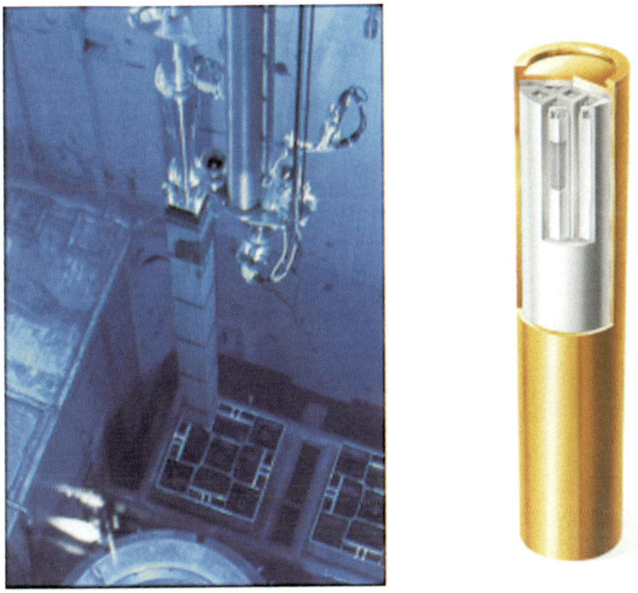

Fig. 2.1. A spent fuel assembly, comprising individual zircalloy-clad elements (tubes containing uranium oxide fuel pellets) being lowered under water into a basket (left) for interim storage prior to encapsulation for disposal in a steel and copper canister (right). Each canister contains several assemblies. Pictures courtesy of SKB, Sweden.

glass-forming material to produce a homogeneous, solid, vitreous waste form. The glass is cast in stainless steel containers that are sealed and may be placed in a further metal container for emplacement in the repository (Fig. 2.2). Alternatives to glass as a HLW matrix have been developed, e.g. the "synthetic rock", SYNROC, and other ceramics, but only borosilicate glasses have been employed at the industrial level.

LL-ILW can come in many forms. It arises principally from reactor operations, from reprocessing SF (e.g. the metal tubes that contained the fuel, and other parts of fuel elements) and from decommissioning nuclear facilities. Long-lived wastes also arise from the production — and the dismantling — of nuclear weapons; this is the primary source of wastes to be disposed of in the WIPP repository in bedded salt in New Mexico, USA. The LL-ILW is generally embedded in a matrix of cement or bitumen inside steel or concrete boxes, to produce monolithic waste packages for disposal (Fig. 2.2).

A wide spectrum of radionuclides may be present in these long-lived wastes. Much of the radioactivity within the majority of waste types can be attributed to the *fission products*, radionuclides that have formed from the fissioning of uranium and other heavy elements in nuclear reactors. Fission products include much of the periodic table of chemical elements and have half-lives that vary from microseconds

Fig. 2.2. Solid radioactive waste forms. Left: Cutaway showing simulated vitrified HLW in a stainless steel production container. The container is 1.3 m high and holds about 150 litres of glass. Right: Cutaway showing simulated cement encapsulated ILW from reprocessing of spent fuel (metallic fuel cladding waste) in a 500-litre stainless steel drum. Several such drums might be emplaced in a steel or concrete box, surrounded by additional cement matrix, for disposal. Pictures courtesy of BNFL.

to millions of years. Those of concern for geological waste disposal have half-lives of a few tens of years or greater. Radionuclides with short half-lives will decay to insignificant levels[4] whilst the waste is in storage during the years immediately after production, prior to disposal. Decay of these short-lived radionuclides is responsible for much of the heat production and intense radioactivity of HLW and SF, which both decline rapidly in the period of some hundreds of years after its production (see Fig. 2.3).

Some categories of LL-ILW contain predominantly *activation products*, radionuclides formed, for example, when construction materials (such as metal components of nuclear reactors) are irradiated by neutrons from fission. Most activation products have short half-lives, but some, for example ^{59}Ni, are long-lived.

The majority of the longest-lived radionuclides in all of the wastes (typically, with half-lives of thousands to thousands of millions of years) are alpha-emitting, *actinide*

[4]A shorthand indication of approaching insignificant levels is after a time equal to about 10 half-lives, when the radioactivity has decayed to less than a thousandth of the original amount: after 20 half-lives it will have decayed to less than a millionth of the original level.

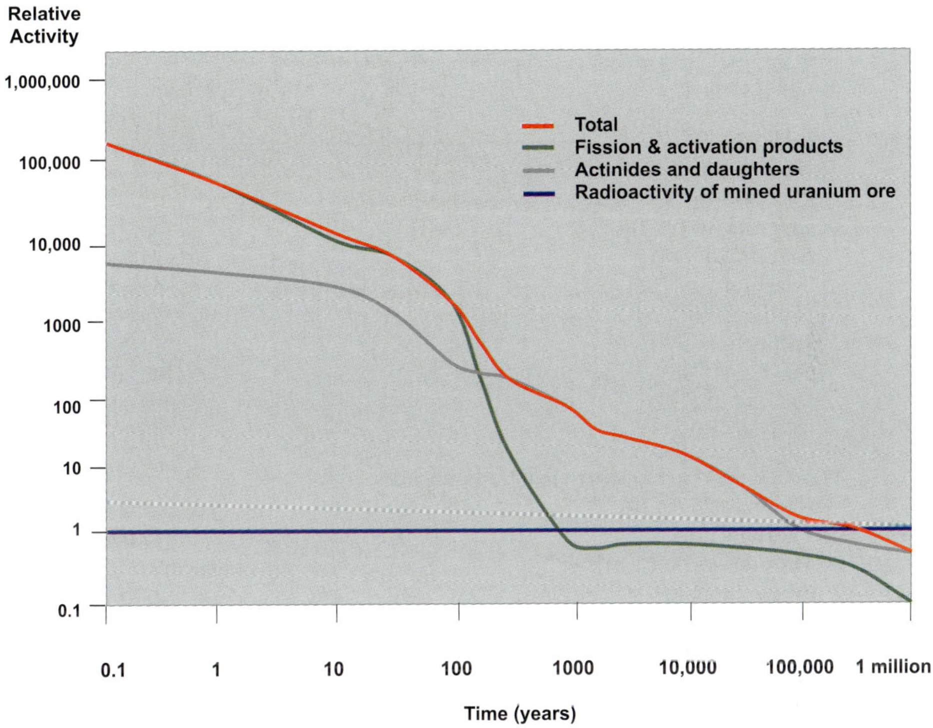

Fig. 2.3. Relative radioactivity of typical spent fuel (a Swedish BWR fuel) as a function of time after discharge from the reactor, showing the early, dominant contribution of the fission and activation products. The sharp plunge in fission product activity between 100 and 1000 years is largely a result of the decay of ^{90}Sr and ^{137}Cs, both with half-lives of about 30 years. After a few hundred years, the actinide elements become dominant. After a few hundred thousand years, the total activity of the fuel is similar to that of the uranium ore from which the fuel was produced (redrawn, after Hedin, 1997).

elements (i.e. with atomic number greater than 89), including members of the natural uranium and thorium decay chains. The *transuranic* actinide elements (those heavier than uranium), such as neptunium, plutonium, americium and curium are formed during the fission process by neutron capture. Table 2.1 shows some of the main, longer-lived radionuclides of interest in managing long-lived radioactive wastes. The above, general description does not cover the origin or nature of all the radionuclides and waste forms that a geological repository might be designed to contain, but it serves to highlight the characteristics of the principal waste groups. Apart from their absolute content of radionuclides and their physical form, a key characteristic affecting safe containment of the wastes is their stability in contact with groundwaters and porewaters in the rock formations and engineered barriers of the repository. Over very long periods of time, such interactions are inevitable,

Table 2.1. Some important radionuclides in long-lived wastes

Radionuclide	Approximate half-life (years)
Fission and activation products	
Carbon-14 (^{14}C)**	5,700
Chlorine-36 (^{36}Cl)	300,000
Nickel-59 (^{59}Ni)	75,000
Selenium-79 (^{79}Se)	65,000
Niobium-94 (^{94}Nb)	20,000
Technetium-99 (^{99}Tc)	210,000
Tin-126 (^{126}Sn)	100,000
Iodine-129 (^{129}I)	16,000,000
Caesium-135 (^{135}Cs)	2,300,000
Actinide and U–Th decay chain radionuclides	
Radium-226 (^{226}Ra)**	1,600
Thorium-230 (^{230}Th)**	77,000
Thorium-232 (^{232}Th)**	14,000,000,000
Protactinium 231 (^{231}Pa)**	33,000
Uranium-234 (^{234}U)**	240,000
Uranium-235 (^{235}U)**	700,000,000
Uranium-238 (^{238}U)**	4,500,000,000
Neptunium-237 (^{237}Np)	2,100,000
Plutonium-239 (^{239}Pu)	24,000
Americium-241 (^{241}Am)	430

**also occur in significant amounts naturally, in rocks, soils and water

as natural deep groundwater conditions become re-established once the repository is closed, and as engineered barriers begin to degrade.

The waste matrices described briefly above are extremely durable and highly insoluble. Under stable, geochemically reducing, low water flow conditions, it would take immense, "geological" periods of time to dissolve them, as evidenced by, for example, the age and stability of uraninite (uranium ore) bodies, analogous in many respects to uranium dioxide fuel.

In fact, the technical concern with containing the wastes permanently and safely results from a relatively small number of fission product radionuclides that are more readily leached from the wastes and which are, more importantly, mobile within deep groundwater systems. Iodine, chlorine and carbon are particularly mobile, and many of the other long-lived fission and activation products in the table below are also relatively mobile in water compared with the actinide radionuclides, which tend to give rise to greatest public concern.

The greater part of the total inventory of radionuclides within any of the wastes discussed above will decay *in situ* within the waste material or the immediately surrounding rock and repository engineered barriers. Small amounts of the more mobile radionuclides can eventually find their way into groundwater and be dispersed in the rock–groundwater system. Some can find their way into the biosphere in extremely dilute concentrations.

The objective of geological disposal is to ensure that as much radioactivity as possible decays without ever reaching the biosphere and that any that could do so would be so dilute as to be of no radiological concern for people or the environment.

Some long-lived radioactive wastes result from the decommissioning and dismantling of nuclear weapons. An increasing amount of such material is expected to require processing over the coming decades. The principal fissile materials used in the construction of nuclear fission weapons are ^{235}U and ^{239}Pu. Disposal of nuclear bombs and warheads does not imply simply burying them in a deep repository. The fissile material would need to be processed so as to make it entirely safe for disposal with respect to any potential for nuclear criticality (see Section 2.5).

2.2 Repository Safety Concepts

Geological disposal facilities for long-lived wastes — if properly sited and constructed — provide passive[5] isolation of radioactive materials. Health impacts, which could, in principle, result both from the radioactivity of the wastes and from other toxic materials that they may contain, are thus avoided. Emplacement in carefully engineered structures buried deep within suitable rock formations is chosen principally for the immense long-term stability that the geological environment provides. At depths of several hundreds of metres, in a tectonically stable location, processes that could disrupt the repository are so slow that the deep rock and groundwater system will remain practically unchanged over hundreds of thousand or even millions of years. This contrasts sharply with the constantly changing, dynamic surface environment. Examples of the relative timescales are discussed in more detail later in this chapter.

The safety of repositories for radioactive waste is based on the multibarrier concept, whereby both engineered and natural barriers within the disposal system act in concert to contain the wastes (Chapman and McKinley, 1987; Savage, 1995). There are two principal components of the multibarrier system:

- the engineered barrier system (EBS), which comprises the solid waste matrix and the various containers and backfills used to immobilise the waste inside the repository excavations;
- the natural barrier, which is principally the rock and groundwater system that isolates the repository and the EBS from the biosphere.

[5]Passive isolation means that, after closure of a waste repository, no further actions are required of future generations in order for the disposal system to provide continued protection.

The extent to which these two principal components act to provide containment, the way in which the different parts of the EBS control the behaviour of individual radionuclides, and the relative weight of natural and engineered barriers at different times in the future evolution of a repository system, constitute what is known as the *safety concept*. The safety concept can be different for each disposal system. Examples of key engineered components of disposal systems currently being considered by national waste management programmes include (see Fig. 2.4):

- **concrete or metal waste containers:** concrete and steel containers, although they may actually last for thousands of years, are generally conservatively assumed in safety analyses not to have any physical containment function after about a thousand years (or even immediately after repository closure in some concepts); however, they can buffer chemical conditions in the repository so as to limit radionuclide release and transport for very much longer times; copper and titanium are expected to have a containment function for up to a hundred thousand years, although in a suitable environment corrosion may take even longer;
- **excavation backfill and buffer (around waste containers) materials:** concretes can limit chemical transport by diffusion for a long period and can also buffer

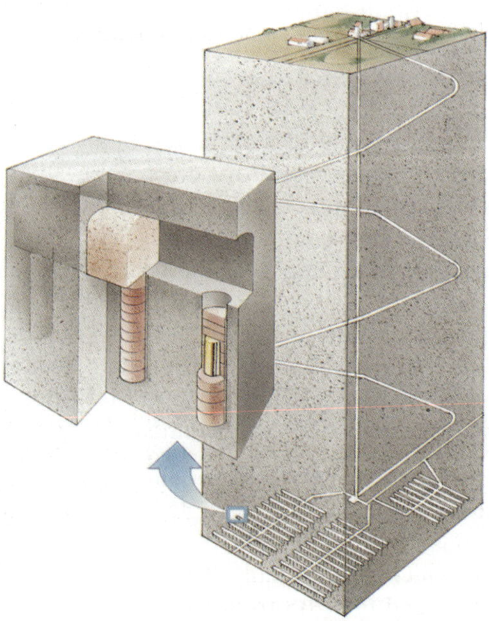

Fig. 2.4. The repository design proposed by SKB (Sweden) for disposal of spent fuel in copper-sheathed steel canisters in a bentonite buffer. The repository would be located at about 500 m depth in hard metamorphic or granitic rocks. A similar design is being developed in Finland. Picture courtesy of SKB, Stockholm.

porewater chemistry and act as a sorbing[6] medium for radionuclides; clays, such as bentonite, are naturally occurring materials which can provide a diffusion barrier for extremely long times.

Some concepts put great emphasis on the protective roles of these EBS materials for protracted periods of time, the longest being Scandinavian concepts for spent fuel disposal in thick copper containers surrounded by a bentonite buffer. Others rely more on the geochemical barriers in the near-field of the repository and on dispersion and dilution in regions of the natural barrier system for some radionuclides. There are significant differences from one national programme to another, from site to site, and from one repository waste inventory to another. In short, the role and weight given to each part of the multibarrier system is very variable.

A central precept of deep geological disposal of long-lived radioactive wastes is that the multibarrier system should work in an integrated fashion to contain the short-lived, highly active radionuclide content of the wastes completely within the EBS until the original radioactivity of these nuclides in the wastes (and the associated heat output) has decayed to insignificant levels. This period is generally in the order of a thousand years (see Fig. 2.3). It is recognised, however, that no EBS design can be relied upon to contain completely the whole inventory, including the long-lived radionuclides, until all radioactivity has decayed to insignificant levels. Examination of Table 2.1 shows that some radionuclides have half-lives of millions of years and more. No engineered system can be guaranteed for such periods, and the containment function must then be provided by the natural barrier, which is stable over geological time periods (millions of years).

What degree of "containment" is actually required in this longer-term context? Figure 2.3 shows that, for spent fuel, containment for a few hundred thousand years would result in a repository that was similar in total activity to a natural uranium ore deposit. This would seem to be a suitable "end-point" for the design of a repository, beyond which the continued functioning of the multibarrier containment need not be of great concern. There is thus a "cross-over point" when the repository can be regarded as having become analogous to parts of the natural environment. In practice, at a well-chosen repository site, the bulk of the remaining radioactivity would be expected to remain immobile, deep within the rock, for millions of years. In the long-term, occurrences of uranium are, of course, never totally "contained" by nature, although many are immobile over immense periods of time. In due course, over millions or billions of years, all ore deposits (including, by analogy,

[6]Sorption is the term given to chemical processes where radionuclides in solution in groundwater become attached to the surfaces of minerals in the rock or the EBS, thus rendering them immobile, unless the chemical environment changes and they can "desorb" and re-enter solution. Some radionuclides sorb much more strongly than others, and some minerals (e.g., clays) have a much higher potential for sorption.

residues of waste repositories) are remobilised and redistributed through the natural environment.

Containment in the intermediate period between the total containment phase of around a thousand years and the "return to nature" state of a few hundred thousand years is thus what primarily concerns the designer when establishing a particular repository safety concept (Chapman, 2002). Contrary to popular misconceptions, the greatest challenge is not containing elements like plutonium since these are extremely immobile in most geological environments. The principal issue is to show that the multibarrier system will limit releases of the mobile radionuclides such as iodine and chlorine. Once water has contacted the waste, it is not possible to exclude the release of some of these radionuclides into the geosphere and, eventually to the biosphere. Safety over the long-term is predicated on these releases to the biosphere being of no consequence because they are very slow and at very low concentrations, controlled by dispersion, retardation and dilution in the rocks and groundwaters of the natural barrier.

In most disposal concepts being considered internationally, the host geological formations (although selected for their low permeability) are still sufficiently permeable that some groundwater movement can occur. The releases are then controlled by the rate and volume of flow of the groundwater. In some clays and claystones, flow is so low that solute diffusion at extremely low rates would be the dominant mechanism for radionuclide migration. A special case is presented by repository concepts involving emplacement in salt deposits. In a normal evolution scenario, no groundwater will contact the waste packages and the predicted releases will be zero at all times considered. The prime focus of safety assessments in clay and salt is then on disruptive processes or events which can disturb the natural barrier.

A carefully chosen geological environment has the potential to act as a cocoon for the repository EBS system, protecting it from gross fluctuations in stress, water flow and hydrochemistry. Large fluctuations in these properties are generally experienced in dynamic regions of Earth's crust, such as near-surface rock and groundwater systems that are more easily and rapidly affected by changes in climate and in land use. The deeper environment is sheltered from these effects, with increasing depth buffering against and smoothing out in time the magnitude of surface perturbations. This isolation from surface effects is an extremely important function of the natural barrier system in the majority of safety concepts, as it enables that part of the disposal system that can actually be designed and optimised (i.e. the EBS) to function predictably for long periods of time.

Figure 2.5 illustrates the concepts described in the above paragraphs in a stylised manner. It can be very difficult to communicate the scales shown on this diagram, which is a much simplified version of many illustrations of risk or hazard as a function of time that emerge from safety assessments of deep disposal. Both the timescale and the hazard scale are logarithmic, a concept that means little to non-technical readers. The fact that most of the hazard occurs in the first thousand years or so, but occupies half of the diagram, while the next 999,000 years of comparatively low hazard occupies the other half, is hard to communicate.

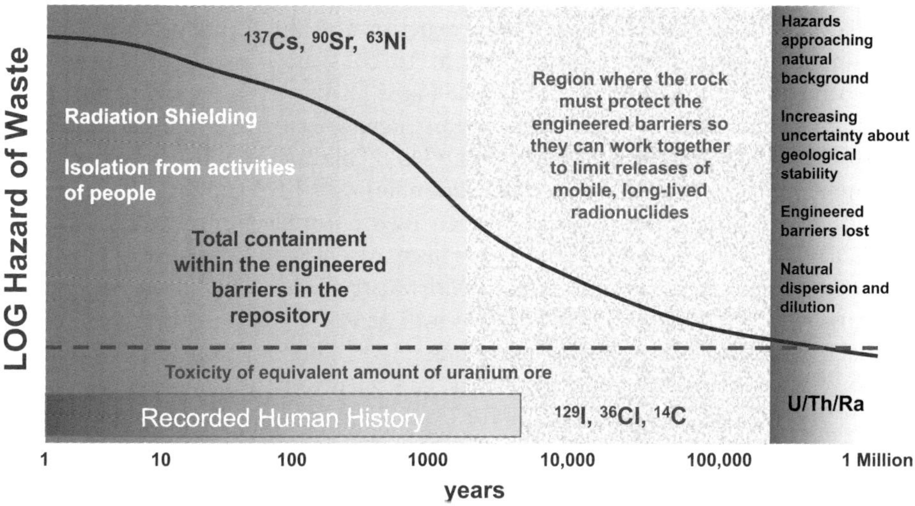

Fig. 2.5. Stylised illustration of how containment in a geological repository works over the first million years after disposal as the hazard, or toxicity, of long-lived waste decreases. The radionuclides shown are typical of those that are of principal concern in safety assessments at different times. The lower box indicates a time period which, if in the past rather than the future, would be equivalent to the duration of the world's recorded history. See text for discussion.

A linear plot would show, on the timescale given, rapid decline in hazard to levels close to natural ores. On the other hand, timescales of even 10,000 years are already too long for the human imagination to grasp readily. As noted in the introduction, standards and regulations have to wrestle with this problem, as they present information to decision makers who may not have a technical background. Some perspectives on the timescale issue are given in Section 2.4.

Suitable geological environments for deep disposal occur throughout the world. Since they can provide the different desirable features mentioned above in different combinations and to different extents, they can vary considerably in their nature. For example, national disposal programmes are considering a wide spectrum of low permeability host rock types, in which it would be feasible to construct and operate repositories at depths of several hundred metres. National choices are dependent to some extent on the local availability of rock formations. These include bedded and dome salt, plastic clay, claystone, mudstone, shale, marl, granite, gneiss, and volcanic tuff.

2.3 Quantifying and Demonstrating Safety

The way in which a repository evolves and its individual components behave in order to provide continuing safety is evaluated using the technique of *performance assessment* (PA). This involves combining experimental and field data with scientific understanding and qualitative observations to construct models of the possible future behaviour of the disposal system (see Fig. 2.6). The function and performance of any individual part of the multibarrier system, or of the complete repository facility, can then be quantitatively analysed. This can assist throughout the development of a waste management programme, with EBS and repository design, and with site selection. The approach often used is to develop a "reference case" model of the expected evolution of the disposal system with time. Best estimates of parameter values can be used, combined with variant cases that explore uncertainty and variability in parameters and the overall sensitivity of system behaviour to particular parameters or model assumptions.

Many near-surface environments will be subjected to considerable changes over the next few hundred thousand years, for example as a result of uplift, erosion and natural, long-term climate cycling combined with more immediate anthropogenic climate change effects. Therefore, PAs need to be able to handle time-dependent processes and their associated uncertainties. They also need to consider the possible future activities of people in the area above and around a repository. These possibilities are generally addressed by constructing scenarios, which describe different possible events, and different magnitudes and interactions of processes. The scenarios are used as the basis for constructing alternative models

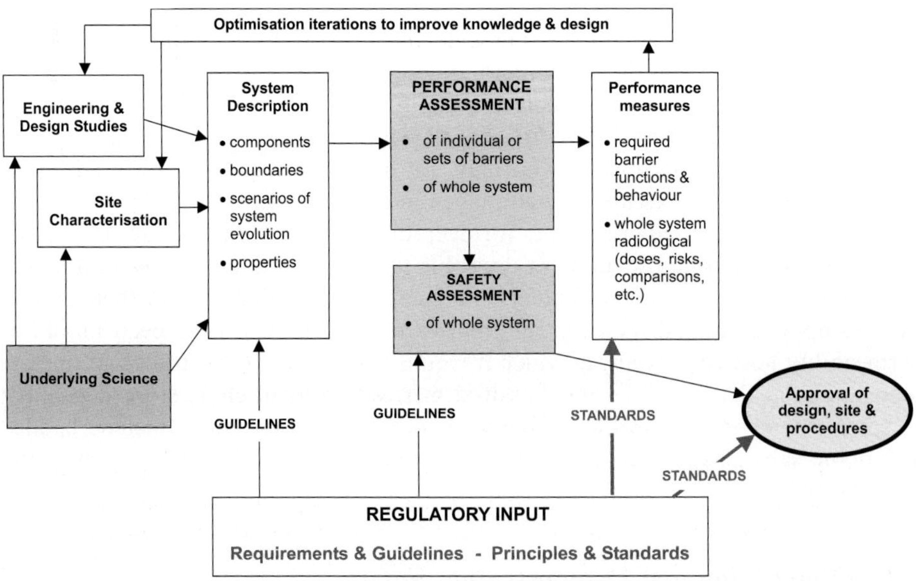

Fig. 2.6. How performance and safety assessments use technical information and are influenced by regulatory requirements. The iterative nature of PA and SA, and how it feeds back to data gathering and system optimisation is also shown.

(to the reference case and its variants) of how the disposal system could respond to these changes.

The results of PA are used to calculate the "fate" of all significant radionuclides in the waste inventory. This would include identification of those that decay *in situ* or in some part of the deep environment, as well as estimation of the rates at which mobile radionuclides might migrate to the biosphere, under a range of circumstances covering uncertainties in future system evolution. These results, when interpreted to show potential radiation doses to people, comprise a *safety assessment* (SA). Clearly, the intention is to design a disposal system at a specific site where it is possible to show convincingly that any such doses would be of no radiological health concern. The results are not, however, intended to give any exact prediction of the future system behaviour; instead, they scope the various potential future developments and allow the analyst and the decision-maker to check that none of these results in unacceptable risks to people or the environment. A typical example of the results of a safety assessment is shown in Fig. 2.7.

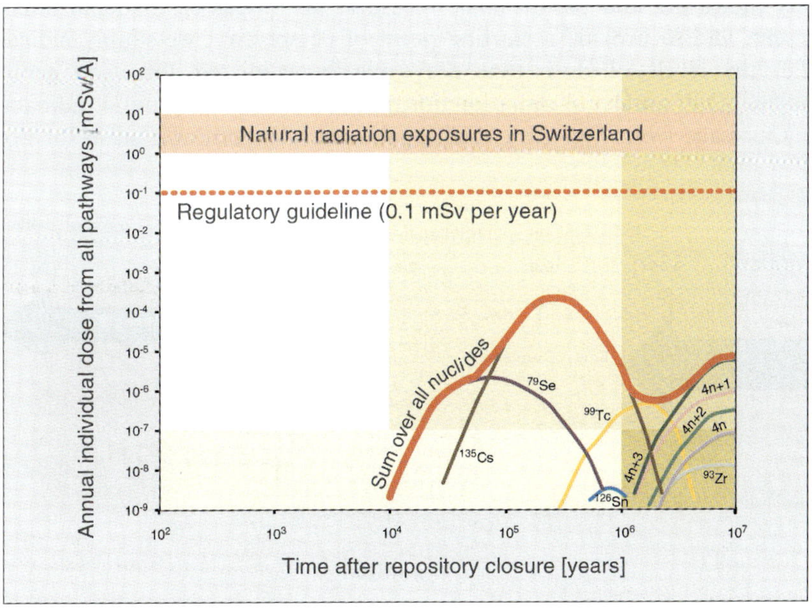

Fig. 2.7. Typical results of a safety assessment and how they are presented, as dose or risk versus time after repository closure. The diagram shows projected individual doses from a hypothetical HLW repository in deep basement rocks in northern Switzerland (Nagra, 1993b). Both dose and time are shown on logarithmic scales and the vertical shaded bars are used to indicate increasing uncertainty in the dose estimates with increasing time, out to 10 million years in this example. Individual curves show the doses attributable to specific radionuclides or members of natural decay series (e.g. 4n + 1). In this case, the highest impact is from ^{135}Cs. The red line shows the total dose from all radionuclides.

The scientific challenge is to understand sufficiently well the physical and chemical behaviour of all system components, their numerous interactions and, most importantly, their long-term behaviour. This task is approached by observations on natural systems, laboratory studies, field experiments and underground testing. It is obviously most straightforward to construct a credible safety case for a simple repository design, located in a highly stable geological environment with simple and easily characterised structure, and this strongly influences siting choices in repository programmes.

What defines an unacceptable level of risk is, of course, a central subject of this review. Risk limits are often laid down in national environmental regulations, which themselves are based on internationally recognised radiological standards. For example, regulations may stipulate target levels of radiation dose (or consequent risk of serious health effects) to hypothetical individual members of the public. A repository developer would be expected to show that any potential doses or risks fell below such target levels, for all reasonable circumstances, and for at least several thousands of years into the future — much longer in some countries.

Safety indicators and performance measures are discussed in detail later. For the present, and to provide a starting point of perspective, we simply indicate, in Fig. 2.8, the levels of dose that are typically estimated for deep geological repositories in the context of radiation doses from natural background radioactivity. The key message from Fig. 2.8 (which again uses a logarithmic scale for dose) is

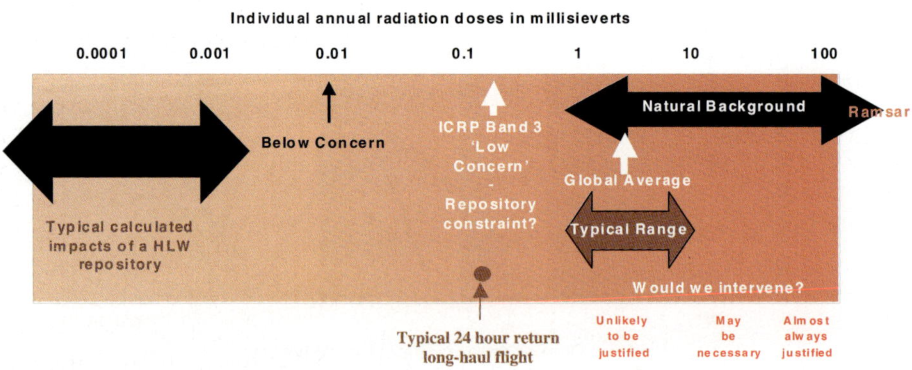

Fig. 2.8. Typical calculated radiation doses from deep geological repositories, compared with doses from natural background (full global range as upper red bar; typical regional range as lower red bar). The highest natural doses (e.g. at Ramsar, in Iran) are ~100 times the average. Also shown are suggested dose limits for a repository (similar to those that arise from a return long-haul flight), radiation levels generally considered as so low as to be below concern, and dose levels from, for example, historically contaminated sites that would be used to decide whether to intervene and rectify the situation. These are discussed in detail later in the book (information mainly from UNSCEAR, 2000, and ICRP, 2000b).

that estimated doses typically calculated from repositories are tens or hundreds of thousands of times lower than both global average natural background doses, or levels from historically contaminated sites at which we might begin to think it worthwhile cleaning up to reduce local doses.

A false conclusion of this description of safety concepts (often reached by sceptics of geological disposal) is that it appears necessary to predict quantitatively in an exact way the behaviour of engineered and natural systems over immense timescales, outside human experience — and to do this with some confidence. As will be discussed in detail later, however, the types of evaluations that are needed are scoping in nature: they cannot, and need not be precise predictions. Nevertheless, a requirement for this sort of long-term analysis is unprecedented in any field of human endeavour, let alone environmental protection. This has raised many issues that need to be accounted for by those assuring this protection: policy makers, setters of standards and regulators in particular. These groups define the requirements, and the implementor, developing a repository programme, needs to know and understand these very well. Accordingly, in the following section we devote some thoughts to this issue.

2.4 The Context of Time

Radioactive waste management is addressing something entirely new by explicitly considering long timescales: there is no precedent upon which to rely either for establishing principles or for judging what people want to see achieved. While other fields of environmental protection and resource planning should be looking at the same problems, sadly, they are lagging far behind.

Throughout the book, and specifically in Chapter 5, we refer continually to time — usually to extremely long times: thousands or millions of years. Outside radioactive waste management, no other field of human endeavour involving engineering, standards and the law does this, as far as we are aware. Some others certainly should, however, since an increasing number of technologies are capable of affecting the future of our planet for extremely long times. A start has been made. Thirty years ago, a few people (the famous "Club of Rome"; Meadows et al., 1972) began to think about the future of Earth's depleting resources, and looked perhaps a hundred years into the future.

Today, a much wider group of scientists is thinking about environmental change and looking forward over a similar timescale. Along with the general increase in interest on sustainability, global warming has made even politicians, perhaps for the first time, look beyond their usual election cycle timespan and think about the relationship between present day policies and societal development over the next decades or centuries. Even for global warming, a reasonably well-defined problem that is widely acknowledged by the public, there is little unanimity of action and considerable political prevarication and procrastination in support of short-term national interests. Politicians are pragmatic, hoping to stay in office by understanding and responding to what concerns people today, i.e. what really motivates us, rather than what we say we believe. Improved living standards for ourselves and for our children are infinitely more important in how we vote and spend our personal money than is some distant threat in the next century.

Expecting serious consideration from any decision-maker of how to respond to environmental issues that have time implications beyond the next few decades might thus appear rather optimistic. The decision of the US government in early 2001 to back out of the draft Kyoto agreements on global greenhouse gas emissions (of which the USA emits 25%), with no proposed alternatives, is a clear example of short-term economic considerations outweighing long-term concerns. Relative to such major potential hazards to future generations, waste management is perhaps a trivial issue. Nonetheless, the waste community has devoted considerable thought in considering how much we should care about hazards from radioactive wastes in the far future. Some of the intellectual efforts are of a pioneering nature and could usefully be transferred to other fields.

Decision-makers, people working in the field and anyone trying to understand and form an opinion on radioactive waste management need to have a framework of comparative timescales in order to grasp the meaning of time, change and impact, and establish an ethical position on standards. The problem is that none of us have a good grasp of time, beyond a few decades. Geologists and astronomers are aware, at least at the intellectual level, of huge timescales. However, few other people have any real conception of what a hundred thousand or a million years involves, so it is hard for them to reach a value judgement about questions involving such timescales (Chapman, 2002).

We use a figure of about 100,000 years frequently in this book to define the order of magnitude of time over which a waste repository (the engineering and the rock surrounding it) would need to function to achieve acceptable containment. The waste containers will remain tight for a few thousand years (or very much longer). Calculations of the performance of repositories estimate that releases of small amounts of radioactivity into the biosphere might begin after some tens of thousands of years, or longer. How can we get a real feel for what all these times mean? The following discussion endeavours to put time into context by comparing these future times with events and developments over comparable times in the immediate past.

The passage of time means much different things in different parts of Earth's surface and crust. Figure 2.9 gives a context for passing time in terms of the processes that occur in three key regions: the human environment, Earth's surface environment, and within stable rock formations deep below the surface. Processes in all of these environments are important when we try to evaluate the future of a waste repository. The first is important because people's activities could interfere with the deep system, the second because surface changes affect the biosphere into which radioactivity may move, as well as having the potential to affect the deeper rock, and the third because this directly controls how the waste and the repository barriers behave.

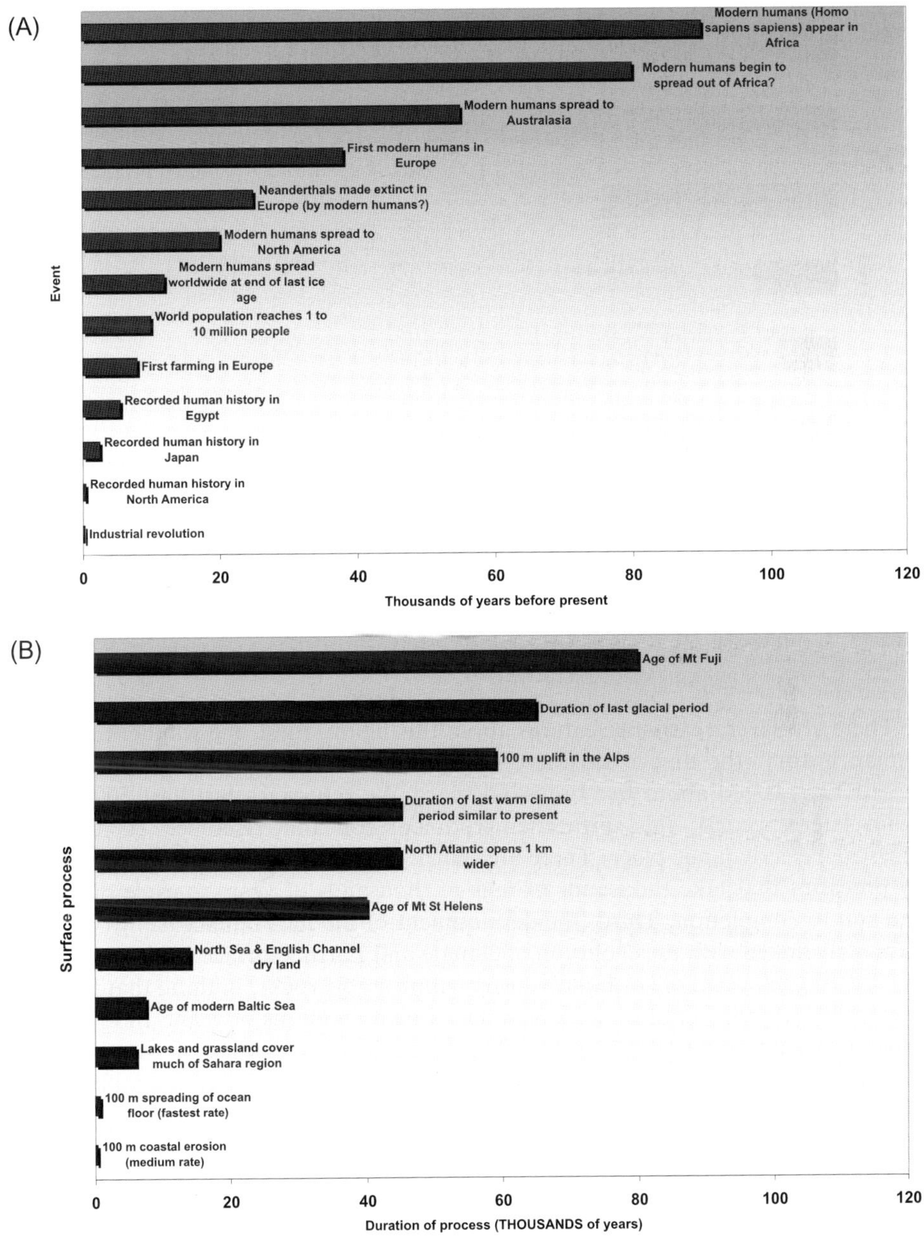

Fig. 2.9. The timescales of processes in three different environments of Earth of relevance to deep geological repositories. (A) the human environment, showing events in the development of modern humans (see text). (B) the surface of Earth. (C) the deep geological environment of Earth's crust where a repository is located. Note the scale of the top two diagrams is thousands of years, while that of the bottom diagram is millions of years.

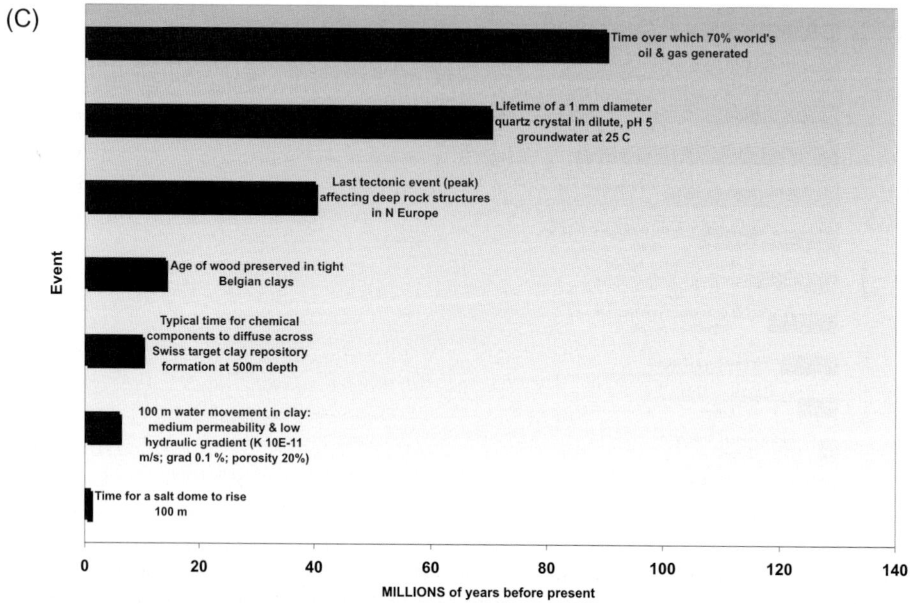

Fig. 2.9. Continued.

There are several striking features about this figure. First, the whole of recorded human history (the time when there have been developing cultures, societies and records) has lasted about five thousand years. We return to this "last 5000 years" again later. Second, the surface environment (on our reference 100,000 year timescale) is a dynamic place. There are sharp, localised events, but these are driven by processes that take thousands or tens of thousands of years to cycle or to start and run to completion. Third, the development of modern human beings has been intimately linked with the evolution of climate and Earth's surface environment over the last 100,000 years or so. Finally, deep within stable rocks, nothing that is readily perceptible happens until we look at times of many millions of years. In many deep geological environments, the physical and chemical properties of the deep rock would be indistinguishable today from the way that they appeared ten million years or more ago. What we are seeing are huge steps between each of these three regions in the "driving timescales" for change: a multiple of twenty or so between the historic period and the climate and surface processes affecting evolution of humans shown in the first and the second diagrams, and a further thousand between these and the deep rock processes in the third.

Returning to our yardstick repository timescale discussed earlier in this chapter, it is interesting to see what has actually happened in the first two environments in the last 100,000 years or so. To recap, this is the period over which engineered barriers will degrade (1000s to 10,000s years), first small releases or radioactivity to the biosphere might occur (10,000s years) and the radioactivity of spent fuel wastes will decay to levels close to natural uranium ore (~100,000 years). There have been two global glaciations in this period, with large regions of the northern hemisphere covered by ice, permafrost or tundra for more than two thirds of the time. This is the period in which modern human beings evolved from early *Homo sapiens* and, according to one widely held interpretation, spread across Earth from an origin in Africa. The spread was controlled considerably by climate change. The last glacial period held back this diaspora in the northern hemisphere until only about 10,000 years ago.

Since then, there has been an enormous explosion in population. Estimates of the "original" population of modern humans in Africa range from a few tens to a few hundreds of thousands of people, about 100,000 years ago. By the start of the great spread into the northern hemisphere ten thousand years ago, the world's population had grown to perhaps one to ten million. Anthropologists speculate that there were "pinch points" during this period when severe climatic events or other causes threatened this small global population with extinction.

By the start of the "last 5000 years" historical period mentioned above, the world population was perhaps as high as that of an average European country today: 10–50 million. By AD 1 it had reached perhaps 200 million. From the start of the industrial period, growth has been explosive. Today, the world holds over 6 billion people. The human status in Earth's environment over 100,000 years is characterised by its steady, almost imperceptible rate of change and tenuous nature during almost the whole period. This is followed by a short period of historical record, the last 5000 years, and an abrupt end to stability a few hundred years ago as the rate of change accelerated massively.

All this has happened within only one tenth of one million years. Natural environmental change has not stopped. It is expected that within the *next* hundred and fifty thousand years, Earth will experience one, possibly two more major glaciations. In many European countries, for example, environmental conditions may remain similar to those of the present only for another 50,000 years, the following hundred thousand years seeing the return of glacial conditions. The last time this happened, only ten or fifteen thousand years ago, there were no modern human beings in northern Europe.

In waste disposal considerations, we obviously need to think differently about processes in the three environments discussed above. The human environment is scientifically intractable and we can make no predictions about it; the surface of Earth is dynamic and we cannot expect detailed evaluation of how it will evolve; the deep environment is comparatively stable and predictable and should form the basis from which we draw confidence in long-term safety.

Different people will gain different perspectives from the discussion above. The following are perhaps some more obvious responses to the time context:

- Society and the near-surface natural environment change so much and on such relatively short timescales that we can't be confident about any projections that depend on predicting such changes. Removal of wastes to the deep underground environment, which changes so little and so slowly, is a logical response to this problem.

- Typical safety assessment calculations indicate that any releases of radioactivity occur several tens of thousands of years into the future. To put this timescale into perspective, a repository would have to have been built perhaps two ice ages ago, by the first modern humans, to be releasing activity to the environment today.
- There have been huge changes in the human state over a few tens of thousands of years and an explosive rate of change over the last few decades, with out-of-control population growth. It is legitimate to ask how much of our scarce present day resources we should devote to protecting hypothetical people in ten thousand years time. We could be accused of developing disproportionate solutions for people who may never exist when we should be concentrating on protecting present day and immediate future generations.
- Releases of radioactivity from a properly designed and constructed waste repository are predicted to be a fraction of a much higher natural radiation background that has existed for the whole of the period during which modern humans successfully evolved against a harsh climate environment and populated the world.
- In the much discussed concept of sustainability, the goal of society should be to leave future generations a natural environment offering the same freedoms and opportunities that we enjoy today. Geological repositories are designed so that they will never represent a hazardous "singularity" in nature, beyond those features (such as ore deposits) that are already present in Earth's shallow crust. Sustainability is thus well-satisfied by a nuclear fuel cycle that includes geological disposal of the wastes, as well as by the reduced dependence on fossil fuels that is more frequently commented upon.

The time context and the issues that it raises provide a good example of how the perception of safety is as much a matter of values and judgements as it is of quantitative science and standards. This is an area where, as noted in the discussion on stakeholders and values in Chapter 1, people's attitudes have not been well tested in the past although some information is now becoming available (see Chapter 5).

2.5 Nuclear Security and Safeguards

As noted in the introduction to this chapter, demonstrating long-term radiological safety is only one part of protecting people from certain categories of long-lived waste. Those that contain fissile radionuclides also present a security challenge. It needs to be shown that these would be practically impossible to recover and extract clandestinely, so that they could be used to make nuclear weapons (Fattah, 2000).

Fissile radionuclides occur in significant quantities in both spent fuel and in wastes produced from decommissioned nuclear weapons, with which we deal first.

2.5.1 *The Global Security Challenge of Dismantled Nuclear Weapons*

With the end of the Cold War, large numbers of nuclear weapons have become surplus to requirements for defence purposes. The reduction in the nuclear weapons stockpiles in the United States and Russia, for strategic as well as economic reasons, is creating large quantities of surplus weapon-grade materials. In May 2002 the United States and the Russian Federation signed a treaty committing each party to reduce its aggregate number of strategic nuclear warheads below 1700–2200 by December 31st 2012. These figures are to be compared with the present world stock of strategic and tactical weapons, which amounts to some 30,000 warheads (95% in the USA and Russia) (Bunn et al., 2002). The stockpiles of separated plutonium and highly enriched uranium (HEU) are estimated at respectively around 450 metric tonnes and over 1700 tonnes. These materials are mostly in the weapons states but civilian plutonium exists also in Belgium, Germany, Japan and Switzerland and HEU at many research facilities in dozens of countries. To make a nuclear weapon requires only a few kilogrammes of fissile material. Accordingly, there are serious concerns that these materials pose a proliferation threat if they fall into the wrong hands.

The danger of theft or diversion or misuse of these nuclear materials has raised the urgency for finding a solution to this international security problem to the highest diplomatic levels worldwide. Currently, the two accepted methods for disposing of these nuclear weapon materials are either to convert them into nuclear reactor fuels or to immobilise them. Both the fissile radionuclides, ^{235}U and ^{239}Pu, can be mixed with much larger amounts of ^{238}U and converted to standard fuel for commercial reactors. Uranium oxide fuel used in power reactors is generally enriched in ^{235}U content (with respect to ^{238}U content) from the natural abundance found in uranium ore (about 0.7%), up to a few percent. Normally this is carried out starting with freshly mined, natural uranium as part of standard fuel fabrication process, mainly using centrifuge technology. Instead, fuel could be made by blending surplus depleted uranium (mainly ^{238}U, of which there is a considerable stock worldwide) with a small amount of ex-weapons ^{235}U to achieve the same result.

Similarly, ^{239}Pu could be incorporated in mixed uranium–plutonium oxide fuel (MOX) now being used in some commercial reactors. The weapons material would then be "burned" in nuclear power reactors and any remaining amount would end up within spent fuel, which would have the same properties as any other spent fuel. Spent reactor fuels are more proliferation-resistant for a longer period of time because of three barriers hindering misuse: their strong radioactivity, the lower concentrations of fissile materials and the less favourable isotopic composition of the plutonium. These all discourage diversion and misuse. Alternatively, fissile radionuclides could be incorporated in very dilute concentrations into the same glass matrix (or a ceramic matrix) used for solidifying HLW. To minimise the problem of long-term nuclear safeguards[7], waste containers would also contain a larger amount of HLW around the fissile material matrix to make the packages so radioactive as to be irrecoverable in practice once they have been placed in a repository (see Fig. 2.10).

Fig. 2.10. The proposed "can-in-can" technique for vitrification of plutonium wastes and production of wastes packages for disposal that meet the "spent fuel standard" for nuclear safeguards (see text). Small quantities of Pu-containing glass (in the four inner "cans") are surrounded by much larger volumes of vitrified HLW. This picture shows a slice across a trial canister containing non-radioactive glass. The overall canister diameter is about 0.6 m. (Courtesy of Savannah River Technology Center, USA.)

This option has only the single proliferation barrier of radioactivity. This makes it harder to meet the "spent fuel standard" (NRC, 2001b) which requires that an appropriate selection of barriers makes the inaccessibility of the plutonium comparable with that of civilian plutonium in aged spent fuel. Whichever disposition method is chosen, both require eventual permanent disposal underground in a geological repository. It has also been suggested that the siting of repositories could be optimised for properly safeguarding these materials, even if this means looking for sites in a worldwide search.

2.5.2 Safeguards for Commercial Spent Fuel

Commercial nuclear electricity generation throughout the world currently results in annual discharges of about 10,000 tons of spent fuel that contains about 1 percent

[7] The term "safeguards" is commonly applied to activities and measures designed to ensure that fissile material cannot be recovered and used to make nuclear weapons again.

plutonium. While this plutonium is not easily separated from the intensely radioactive spent fuel and is not of the same quality as plutonium removed from weapons, it can still be used as a threat or in a crude weapon. Smaller quantities of highly enriched uranium form the fuel for research reactors around the world. The amounts of these nuclear materials will increase if nuclear power expands to meet the growing demand for energy in developing nations and as a contribution to reducing greenhouse gases.

Since the radioactivity of spent fuel decays over time (with most of the fission product activity decaying before that of plutonium), it loses its natural proliferation protection, so long-term storage is not a permanent solution to assure security. Additionally, some spent fuel could be produced with inferior barriers to diversion through lower burn-up rates or unplanned early removal from reactors. Today, most commercial spent fuel worldwide is maintained under a strict safeguards regime and thus presents no urgent security threat. Ultimately the material must be made as inaccessible as possible. An obvious approach is to gather the fuel to a central location that can be closely guarded and to keep it in the most secure facility possible. Obviously, a deep geological repository can be made very secure.

2.5.3 Increasing Global Security with National and International Repositories

For those nations with good prospects for successful national repository programmes, there are very useful non-proliferation advantages to concentrating materials from numerous locations into a carefully selected national site that is technically easier to safeguard. This security argument gives good reason for avoiding unnecessary delays in implementing national repositories. However, at least 33 nations currently have commercial nuclear power programmes and others have research reactor programmes. It is unlikely that every one of these countries (particularly the smaller programmes) will possess the political, economic, and geological factors necessary to create permanent deep repository disposal programmes for their materials, soon or ever.

In a future of increasing quantities of excess fissile materials from dismantling nuclear weapons and commercial nuclear industry activities, there is a convincing requirement for creating new international spent fuel storage facilities and disposal facilities to help in solving this security threat. A global system of a few disposal facilities in carefully selected, isolated areas under multinational scrutiny should be preferable to many small national facilities that may be located in less than ideal conditions. Using a global approach, safe and secure permanent geological disposal sites for spent fuel and immobilised nuclear weapon materials can be identified in potential host nations possessing internationally acceptable political and non-proliferation credentials. Participating nations would open their nuclear programmes to even greater transparency. By accepting stringent, continuing oversight by the international community (even more international in nature than current effective IAEA non-proliferation and safety activities) a host nation could significantly contribute to the essential climate of public trust that must exist if a multinational solution is to be acceptable.

The sites can be selected at remote, easily monitored locations that would simplify detection of diversion attempts. Design, construction, and operation of international repository sites can be optimised for safeguards considerations, to maximise their non-proliferation and security aspects. Repositories are a cost-effective means of enhancing safeguards and an added benefit would be the reduction of monitoring costs of the IAEA, whose safeguards responsibilities have increased in recent years without a corresponding budget increase.

Governments are seeking innovative, flexible, commercially sustainable solutions that can supplement current diplomatic efforts to ensure that fissile materials are kept under control. Assured storage and disposal routes in support of arms control objectives are essential. Storage schemes in Russia have been proposed (e.g. Cochrane and Paine, 1998). A national or international repository that accepts ex-weapons materials may encourage faster conversion of nuclear weapon materials into nuclear fuel to achieve the spent fuel standard that is the goal of non-proliferation programmes. The advantages of multinational repositories in this respect have been emphasised, e.g. by Pellaud and McCombie (2000) and Stoll and McCombie (2001). A comprehensive study of the issue is contained in the recent book "Megatons and Megawatts" (Garwin and Charpak, 2002).

Chapter 3

Ethics

Within the radioactive waste management community much time and effort has been devoted over the years to debating ethical issues underlying the concepts developed for safely handling and disposing of long-lived wastes. Many regulatory regimes governing disposal explicitly acknowledge the ethical principles involved and attempt to base their requirements on these principles. After an introductory overview, this chapter identifies the main principles, discusses their relevance and derives "messages" which should influence the development of safety criteria for deep geological repositories.

3.1 Early Ethical Considerations

In the early years of radioactive waste disposal studies, the problem was primarily regarded as a technical and economic challenge without much explicit recognition of political, social and ethical aspects. There was, nonetheless, direct recognition of the key importance of ensuring the safety of people and the environment. In 1955, the guidelines for the US National Academy Committee on Geological Aspects of Radioactive Waste Disposal already included the following principles (see NRC, 1966):

- Safety is a primary concern, taking precedence over cost.
- Radioactive waste, if disposed of underground, should be isolated as permanently as possible from contact with living organisms.

Subsequent work concentrated on the technical issues, although, at one of the earliest International Conferences, in Otaniemi, Finland, in 1979 (IAEA, 1980), the protection of future generations from risks was addressed explicitly by Jauho and Silvennoinen, who recorded that:

> ...guarantee arrangements seem to be necessary in order to prevent society from passing on excessive risks from nuclear wastes to future generations.

Emphasis was then on economic risks to future generations, but in the eighties, explicit attention was paid to ethical issues during development of objectives and principles for radioactive waste management by the OECD Nuclear Energy Agency (NEA) and the IAEA (NEA, 1984b; IAEA, 1989).

The NEA report concentrates on how to apply operational radiation protection principles to practices which might give doses only in the far future. The ethical basis behind such considerations is reflected in the report's statement that:

> ...the reasons for adopting the same principles when dealing with hypothetical exposures to the public in the far future from today's waste disposal practices are a desire for equity, in that future generations should be given the same degree of protection that is given to the present generation.

The IAEA Principles (IAEA, 1989) were much broader, reflecting various ethical aspects of waste disposal. They were reformulated, after much international discussion, to give the wording contained in *The Principles of Radioactive Waste Management* (IAEA, 1995b), extracts from which are included in the following section.

3.2 Ethical Principles in IAEA Documentation

IAEA (1995b) contains the following ethical principles protecting current and future generations:

- Principle 3: Protection beyond national borders: Radioactive waste shall be managed in such a way as to assure that possible effects on human health and the environment beyond national borders will also be taken into account.
- Principle 4: Protection of future generations: Radioactive waste shall be managed in a way that predicted impacts on the health of future generations will not be greater than relevant levels of impact that are acceptable today.
- Principle 5: Burdens on future generations: Radioactive waste shall be managed in a way that will not impose undue burdens on future generations.

The Safety Principles of the IAEA have formed a basis for the major IAEA Joint Convention on the Safety of Spent Fuel Management and on the Safety of Radioactive Waste Management (IAEA, 1997a) which entered into force in 2001. The convention contains the following relevant articles.

- ART 1(ii) to ensure that during all stages of spent fuel and radioactive waste management there are effective defences against potential hazards so that individuals, society and the environment are protected from harmful effects of ionising radiation, now and in the future, in such a way that the needs and aspirations of the present generation are met without compromising the ability of future generations to meet their needs and aspirations;
- ART 4 (vi) strive to avoid actions that impose reasonably predictable impacts on future generations greater than those permitted for the current generation;
- ART 4 (vii) aim to avoid imposing undue burdens on future generations.

For waste disposal, it is important to note that the requirement is not that the **actual** burdens on future generations are less than those imposed on current populations. The common designs for repositories are such that **no** releases are to be expected for very long times into the future. Hence even very small releases at far future times can lead to radiation exposure which although small are higher than those for current generations. The wording of the IAEA texts is carefully chosen to indicate that the only practicable measuring stick is those doses deemed acceptable to today's society.

3.3 Ethical Discussions within the OECD Nuclear Energy Agency

A further, important international document is the *Collective Opinion on the Environmental and Ethical Basis of Geological Disposal* produced by the NEA/IAEA/EEC in 1995 (NEA, 1995a). This consensus view, drafted following a 2-day, wide-ranging workshop on Environmental Aspects of Long-Lived Radioactive Waste Disposal (NEA, 1994), is commented upon in more detail below. The workshop included also experts from outside the radiation protection or the radioactive waste management fields — but did not include representatives of NGOs or members of the public.

Over the years the OECD/NEA has issued selected "collective opinions" intended to record the views of its senior committee of experts on key waste management issues. The sequence of these opinions is somewhat paradoxical, and directly indicative of the defensive battle being fought by the nuclear community. A first paper gave the consensus view that radioactive waste disposal **could be carried out safely** (NEA, 1985). A second recorded and justified the consensus view that **adequate methods were available for assessing repository safety** (NEA, 1991) a conclusion, which should obviously have been drawn prior to the first. The most recent should, in fact, have preceded both the others, in that the documented consensus is that the concept of geological waste disposal **rests on a firm ethical basis** (NEA, 1995a).

The collective opinion on ethical aspects was carefully prepared following the previously mentioned workshop on the topic. For the workshop, a background document was prepared listing, in an open manner, the numerous issues to be tackled, and posing the direct question as to whether disposal concepts fit into the framework of sustainable development and ethical responsibility which is accepted today.

In the background text to the Collective Opinion, attention is focussed upon:

> ... the achievement of intergenerational equity by choosing technologies and strategies which minimise the resource and risk burdens passed to future generations and it is recognised that each generation leaves a heritage to posterity involving a mix of burdens and benefits and that today's decisions may foreclose options or open new horizons for the future.

A set of guiding ethical principles is developed in the NEA document; these are broadly similar to the principles of the IAEA mentioned above. Two issues, however, are more strongly emphasised. One is that:

> ... a waste management strategy should not be based on a presumption of a stable societal structure for the indefinite future, nor of technological advance.

This principle leads to rejection of indefinite storage strategies requiring continuing deployment of resources in favour of geological disposal concepts offering permanent protection. The second issue, discussed more extensively in the Collective Opinion, is the wish to ensure that one does:

> ... not unduly restrict the freedom of choice of future generations.

As described below, this ethical principle has gained in prominence in the past several years. Since taking any specific course of technical actions tends to exclude or make more difficult other options, the approach often suggested for maintaining freedom of choice is to ensure that any actions are reversible. In waste disposal, it is judged that an incremental process, involving development of deep repositories in a stepwise fashion over decades, meets this requirement — even when disposal facilities have no deliberate provisions for waste retrieval following repository closure.

In its summary collective opinion on the ethical aspects of waste disposal, the Radioactive Waste Management Committee of the NEA considers that responsibilities to future generations are best discharged by the strategy of geological disposal and believes that both inter- and intragenerational issues are thereby taken into account. Intragenerational equity is used here to mean, broadly, fairness across current generations, with intergenerational equity being understood as fairness towards future generations. Ethical responsibilities to current generations require, for example, that we should keep in perspective resource deployment in all areas where there is potential for reduction of risks to humans and that the implementation of geological disposal should proceed stepwise with ample opportunity for proper public participation in the decision process.

3.4 National Positions on Ethical Issues

At a national level, there have also been numerous position papers on ethical issues. In Sweden, for example, the advisory council, KASAM, organised a symposium on the subject in 1987 (KASAM, 1988). KASAM was the first organisation to place strong emphasis on the overriding importance of the above-mentioned principle of keeping future options open. Other countries have addressed the ethics issue less formally or publicly. In Canada, specific studies have been done to give ethical input to the national strategy for disposal of spent fuel (see Roots, 1994). In Switzerland, as a preliminary to revision of the government regulations governing long-term disposal of radioactive wastes, a seminar was held at which ethical issues were presented by experts from outside the nuclear community. The USA has an extensive literature on the general question of achieving equity between successive generations and this discussion has been taken up by those concerned with radioactive waste management (e.g. Schrader-Frechette, 1994; Okrent, 1994).

In recent times, the issue raised by KASAM of maximising the freedom of choice of future generations has led to much discussion on the potential conflict between this ethical view and the principle of minimising future burdens. It is the conviction of a significant body of persons that imposing burdens on future generations is, in fact, **more ethical** than restricting their freedom of choice. This point has been made in the following way (Miller, 1998):

> The burden of an imposed responsibility may be a lesser evil if the alternative is to inflict a threat of harm with no possibility of mitigating the harm. The monitoring system and retrieval option give future generations options they might not otherwise have.

This body of opinion includes the vocal lobby calling for indefinite surface storage of radioactive wastes as an alternative to implementing deep repositories which they regard as potentially harmful. Included amongst the supporters of this view are not only declared opponents of nuclear power (such as those propagating the "guardianship" concept[8]). There is in fact, sometimes a kind of "unholy alliance" between such nuclear opponents and some strong nuclear promoters who also argue for very long surface storage. In the latter case, however, a strong additional argument is that extended surface storage postpones expensive geological repository construction. A recent example of such unbalanced arguments and their refutal is contained in exchanges between Cave (2001) and McKinley and McCombie (2002). Long-term storage, however, is also advocated by neutral bodies such as the KASAM organisation or the Dutch Government, which has recently ruled that **all** toxic wastes must be stored in a retrievable fashion. In many national waste management programmes, a compromise is aimed at, in which wastes will be emplaced in such a way that they will be safe for all time without any further efforts being required from future generations — but which will enable their retrieval should future generations decide to recover the wastes. The following chapter enters into more detail on the issue of reversibility or retrievability since these are points which are today directly affecting the proposed designs and operational concepts for geological repositories.

The remaining discussion in the current chapter aims at a structured approach linking ethical principles to specific requirements on disposal programmes and thereafter to safety and other criteria established in national programmes. The fundamental principles are **fairness** or **equity** for current and future generations; these two concepts, as mentioned above, are labelled respectively **intragenerational** and **intergenerational** equity. They are treated separately below.

---

[8]"Guardianship" (NGL, 1992) is promoted as a concept in which wastes are deliberately kept on the surface in highly visible engineered structures. The arguments made are that this will focus attention on their proper care and maintenance and that they will act as a "monument to a failed technology", nuclear power.

3.5 Intragenerational Equity Aspects

Intragenerational equity means that within current generations it is important to ensure that our finite resources are spent sensibly on solving environmental problems, taking into account the relative scale of the potential impacts and also the spatial distribution of risks and benefits. It implies also that decisions on how to achieve these aims are made in a fair and open manner, involving all sections of society. In the following, we address a series of intragenerational equity issues and try to derive from this the messages which are valuable for waste disposal implementors or regulators.

3.5.1 Health Risks to Current Populations

The ICRP has an initial principle of radiation protection which holds that any practice leading to radiation exposures to populations must be justified (see Chapter 6). For waste disposal, the practice is usually taken to be part of the larger issue of nuclear power production (or other nuclear technology application), so that explicit justification of disposal in this sense has not been an issue. The criteria set for allowable exposures to current populations from operational activities is also not a disposal-specific issue since the relevant facilities and activities are treated like any other nuclear application.

The goal of ensuring that current populations are not exposed to unacceptable risk from radioactive wastes can be met without implementing disposal. Decades of practical experience have shown that interim storage facilities can be built, operated and maintained in such a way as to present no significant public hazards. This lack of urgency for disposal facilities is one reason why there has been no strong political or public pressure for their implementation. Recent concerns about malevolent acts of terrorism, as will be discussed later, may in the near future prove to be a stronger driver for getting hazardous materials underground.

In radiation protection in general, ethical considerations would argue that intragenerational equity would require the levels of risk criteria to be set relative to other activities that are potentially hazardous to the public. In fact, only few countries have a uniform regulatory framework that should encourage this; examples of countries that do are the USA with the Environmental Protection Agency and the UK with its Environment Agency. Even in these organisations, although there is much talk of risk-informed regulation, there is no real pressure to use uniform risk criteria. The widely recognised "nuclear dread" factor associated with radioactivity tends to lead to especially strict formulation and enforcement of regulations in the nuclear area, including waste management. In Chapter 13, we go into more detail on the specific comparisons between regulations for radioactive materials and other toxic materials.

3.5.2 *Social and Economic Impacts*

Despite strict regulation of radiation exposures, there is an additional ICRP requirement to maintain exposures "as low as reasonably achievable, social and economic factors being taken into account" (ALARA). On the one hand, the economic part can justify arguments against exorbitantly expensive measures (e.g. overdesign of engineered barriers which do not greatly increase safety). On the other hand, the social argument can justify fully weighting also the subjective arguments of the public — and hence being prepared, for example, to spend more resources per life saved on nuclear than on conventional risk reduction measures. In practice, the resources which society currently invests to save human lives vary enormously, from millions of dollars per life saved for the expensive measures taken to protect people from low-level radiation, down to mere tens of dollars for saving lives by immunisation programmes.

3.5.3 *Spatial Distribution of Burdens and Benefits*

At an international level, the IAEA Principle 3 on "protection beyond national borders" addresses the geographical distribution of negative impacts. The IAEA also has guidance on international transfers in its Spent Fuel and Waste Convention and on transboundary effects in its Principles. The ethical rules proposed do not, it should be stressed, exclude transfer of wastes between sovereign States. In practice, this happens regularly for chemotoxic wastes and has happened often in the past also for radioactive wastes (O'Neill, 2000, 2002). For example, the reprocessing nations France and the UK originally accepted that they would dispose of the resulting wastes along with their own national waste inventories.

Spent radioactive sources are expected to be disposed of by the country that buys them. The IAEA has specifically studied the conditions that should be fulfilled for multinational waste repositories (IAEA, 1998a, 2003) and the EU has debated equivalence principles for waste substitution. More recently, however, there have been marked movements towards limiting or banning transfer of wastes. For example, countries like France, Sweden, Finland, and Russia have banned waste imports, although this last country is currently trying to amend its legislation to allow import. The reprocessing countries France and the UK now insist on returning wastes to customer countries. The UK has adopted a policy of "self-sufficiency" in this area. There have also been assertions that transfer of radioactive wastes is somehow morally unjustified. In practice, there are no ethical reasons for treating radioactive wastes differently from other commodities, including chemotoxic wastes. There are, of course, strong ethical reasons for not exporting hazardous wastes to any country that does not have the appropriate technological and societal structures to ensure that these wastes are properly handled. The arguments against waste transfers in the case of willing and capable host nations being prepared to accept waste imports are less a matter of principle and more of political expediency. In waste management circles, at least, the common view is that international repositories will eventually be implemented — although some believe that national facilities must first "show the way" (NRC, 2001a). Recently, the European Union has recognised the potential advantages of regional repositories (CEC, 2002). Both authors of this book are firmly convinced that multinational or regional repositories are a necessity and both have been involved directly in efforts to encourage such developments (McCombie and Stoll, 2002; Black and Chapman, 2001).

At a national level, the distribution of burdens and benefits is a key issue in the siting of waste repositories. Today, it is a widely accepted practice that a host community should be compensated for its willingness to accept a common facility which is for the good of a wider population. Specific negotiations on such issues have taken place in numerous countries, including Canada, Finland, France, Japan, Sweden, Switzerland, Taiwan and the USA (Richardson, 1998). The benefits offered are properly regarded as fair compensation for the disruption involved in the work and for the willingness to provide a service to society, rather than as bribes or as risk premiums. Success has been varied. Local communities have reached agreement with repository proponents in Switzerland at the Wellenberg site (which was subsequently dropped for other reasons) and in Finland at Olkiluoto; however, the situation earlier in the USA and more recently in South Korea has been that incentives offered, even for low-level waste siting or for interim storage, have not been taken up. In Japan, the issue is open (NUMO, 2002a).

3.5.4 Public Involvement

Intragenerational equity requires that the public be given open access to information, that their concerns are appropriately weighted and that they can participate in the relevant decision-making processes. In many countries today, information on waste management is freely available; the advent of the world wide web has also made this information easily accessible. This position has been reached despite the initial tendency to secrecy, bred in nuclear weapons programmes and taken over into commercial power activities. Increasingly, there is also a trend towards engaging the public in the debate and ultimately in the decision processes. This is sometimes done informally with public fora or public enquiries. In some cases, e.g. in the rule making of the USA, there is a highly formalised mechanism for gathering public comments on key issues. The recognition that public involvement in waste management issues is vital has grown also within the waste management community. A recent example of this is the establishment by the OECD/NEA of a forum on stakeholder confidence dedicated explicitly to this topic (NEA, 2000a).

The ultimate instrument of public participation is perhaps that of a referendum in which every person can record his opinion. In some countries (e.g. Switzerland), referenda are usually binding on the responsible political authorities. Even in countries where this is not the case, there is an apparent trend towards implementors committing themselves to abide by the results of consultative referenda (e.g. in Sweden). An important caveat, which is often forgotten here, is that the public cannot be expected to master all of the technical issues involved, so that the implementor and regulator have a direct responsibility to make as clear as possible the scientific issues on which there is a broad consensus. A further key point is that increased access does not of itself guarantee more democratic outcomes. Powerful lobbying groups on either side of the debate can and do work towards reaching goals shared only by their minority group of peers.

3.6 Intergenerational Equity Aspects

Intragenerational equity involves ensuring fairness towards future generations; it is directly related to the topical subject of sustainability. The basic tenets are that we do not pass on burdens unnecessarily and that we leave future generations with the same freedoms and choices that we have. In the following, we address intergenerational equity issues and try to derive from these the messages that are valuable for waste disposal implementors or regulators.

3.6.1 Risks to Future Generations

The IAEA Principles maintain that future generations should not be exposed to higher risks than current generations accept. This would lead to dose or risk criteria for future exposures being set equivalent to those for operating facilities. In practice, the argument has sometimes been made, e.g. in the Swiss Regulation R21 (HSK & KSA, 1993), that since the current generation is the beneficiary of nuclear power future doses should be less. This has resulted in dose limits like 0.1 mSv/a being set for the future, whilst current radiation protection limits are significantly higher.

The ethical arguments concerning protection of future generations have not gone unchallenged, however (e.g. Okrent, 1994). Firstly, it has been pointed out that future generations do indeed benefit from nuclear technology through the technical advances made, the conservation of fossil reserves, the reduction in greenhouse gases, etc. More philosophically, it is pointed out that not discounting future risks leads to positions that do not reflect how society actually functions. For instance, allocating equal values to all future lives would cause society to invest in measures which might save 100 lives 100 years from now rather than in measures to save 99 lives next year. In practice, it has been shown that people do discount future risks and indeed that the discount rates assumed are roughly the same as those used for financial discounting (Ahearne, 2000).

In addition, the inability to guarantee long-term or effectively permanent institutional control over long-lived uranium mining wastes disposed of at the earth's surface or over historical "legacy wastes" in countries where defence programmes have resulted in large-scale contamination, means that we are implicitly accepting (for this type of waste, and some NORM[9] wastes) that future generations may have lower levels of protection than today. This is causing re-examination of

[9]Naturally Occurring Radioactive Materials, such as those arising from uranium mining.

the appropriate balance of radiological protection standards for the future for these materials. The most commonly accepted principle today for disposal of nuclear fuel cycle wastes is that future generations must be protected for very long times (at least 10,000 years) to at least reach the level of protection expected by today's generations; for extremely long times the growing tendency is to then make comparisons with natural sources of radiation, such as ore bodies (see Section 2.2).

3.6.2 Burdens and Benefits for Future Generations

The potential burdens on future generations do not involve only radiation risks. The most obvious other risk is financial, and this is discussed separately below. In any ethical discussion on future impacts of waste disposal, one should also address the benefits that can result. The most obvious benefits associated with HLW and spent fuel are related to the overall practice of nuclear power — and hence subject to controversial discussions.

However, serious debate on ethics must acknowledge also the potential benefits of technology advances and increased energy availability. For nuclear power, additional arguments are conservation of fossil reserves and reduction of greenhouse gases. Outside the scope of power production, other applications of nuclear technology in medicine, research and industry certainly produce benefits for current and future generations — but these are often under-emphasised by both proponents and opponents. The huge importance of these points for all future generations should be more strongly stressed in debates on the ethics of nuclear power and radioactive waste disposal.

3.6.3 Financial Risks to Future Generations

Implementing repositories will be expensive and postponing this task for long times means that these costs will fall on future generations. For this reason, serious waste management programmes set aside funds to cover these future liabilities. The pioneering example here was Sweden where a fund, fully segregated from the utilities and from Government, was established early in the programme. Many other countries now have similar funds, although these are sometimes open to appropriation by Governments for other uses, as in the USA, or are left within the utilities, as was the case until recently in Switzerland.

A complication which can arise, even when funds are properly regulated is that there is a large uncertainty on the extent of these future liabilities. No high-level waste repositories have been implemented to date and cost estimates for the facilities tend to rise monotonically with time. Accordingly, there is continuing debate about the necessary level of the funds. This is aggravated by the fact that the necessary rate of accumulation of funds depends upon the scenarios assumed. If one allows for premature closure of power plants or for malfunctioning of repositories, leading to a need for retrieval of the wastes, this would obviously require more funding to be set aside. Despite these open questions, the fact that the full costs of all waste management liabilities are internalised in nuclear electricity prices in many countries is often ignored in the debate on nuclear power. Should the same requirements eventually be placed on other types of power plants (e.g. fossil plants producing CO_2), the economics of these could change rapidly.

3.6.4 Maximising Freedom of Choice

The issue of not unnecessarily restricting the choices of future generations was originally highlighted in Sweden. This aim can obviously cause conflict with the principle of minimising potential burdens and also with the issue of nuclear safeguards. In the extreme case, **all** choices can be left open by current generations postponing all decisions on waste management. Wastes would not be conditioned, in case better methods become available; disposal would not be implemented in case alternatives like transmutation provide perfect solutions; repositories would not be sealed in case we wish to retrieve the wastes with ease; etc. This approach, however, also passes on all burdens and is certainly not ethical.

In practice, there is a strong, and increasing, tendency in the area of waste disposal to try to provide a compromise. Implementors are trying to develop repositories which provide future safety but also retain options for change. Retrievability of wastes has become a major topic (see, for example, IAEA, 2000) and is discussed in depth, later in this review. Explicit studies have been performed to illustrate that recovery of wastes from any geological disposal facility is in principle possible, if enough time and money are invested. Trial disposal for decades with potential recovery thereafter is foreseen in Sweden. Long periods of open tunnels which ease retrievability are planned in, for example, the USA (at Yucca Mountain) and were proposed in Switzerland (for the L/ILW repository at Wellenberg). Such approaches need to ensure that components of the engineered barrier system vital for long-term safety do not degrade during the open period.

It is also becoming common to require comprehensive measures for archiving of repository data in order to facilitate potential retrieval in the far future. Already during the long phase of preparation and operation of repositories, the recommended approach is to progress in a stepwise fashion that preserves reversibility of each step for as long as possible. People do not like to make irreversible decisions if these are avoidable. In the ethical debate surrounding disposal, achieving the correct balance between maximising freedom to change direction and minimising future burdens is perhaps the most sensitive of all current issues.

3.7 Other Ethical Principles

3.7.1 Sustainability

The topical issue of sustainability is closely related to intergenerational equity. The most widely accepted definition of "sustainable development" is that of the

Brundtland Commission, "*development that meets the needs of the present without compromising the ability of future generations to meet their own needs*" (World Commission on Environment and Development, 1987). Most of the relevant points for waste disposal have been touched on above in the discussions on burdens and benefits. Nuclear power — with properly implemented, safe disposal — is sustainable since it contributes to reducing hazards in the human environment, conserving hydrocarbon resources, etc. Specific repository siting measures can be taken to enhance such attributes; for example locating repositories in areas where intensive human usage is unlikely and where no restrictions are put on the availability of natural resources.

A recent development in waste disposal, which pushes the issue of preservation of the environment yet further, is the increased emphasis on direct protection of organisms other than man. The earlier assumption that protecting man automatically protects other species has been questioned and appropriate approaches are being discussed in some programmes (IAEA, 1999b).

3.7.2 Precautionary Principle

This principle calls upon society to take prudent preventative actions to deal with risks with potentially very serious consequences, even if there are doubts and scientific controversy surrounding the evidence. A topical non-nuclear example concerns global warming. Although a scientific debate on the mechanisms, rate and consequences of warming still persists, governments have become sufficiently convinced of the potential for disastrous consequences that they are willing to try to reduce CO_2 emissions now. Whilst the precautionary concept is obviously laudable, its implementation can easily lead to misuse of society's resources in a manner which conflicts with the principle of intragenerational equity. In general, overemphasis on the precautionary principle can lead to a luddite-like rejection of new technologies, and hence its application calls for sound judgement. For deep geological repositories, it can generally be argued that any future impacts will be localised, not of a catastrophic nature and not irreversible, so that the precautionary principle has limited relevance.

3.7.3 Polluter Pays Principle

The fact that polluters should not be subsidised or, put more constructively, that waste management costs of industries should be internalised, is widely accepted and influences environmental legislation in almost all countries. Difficulties can arise in assessing the costs, in particular of pollution which is diluted and dispersed (e.g. CO_2 emissions). Nuclear power and geological disposal are more straightforward and, as described above, mechanisms to ensure costs are in place in most countries. The more generalised form in which "users pay full costs" is more difficult because the costs of avoiding pollution are relatively well-defined compared with the costs of, for example, using up natural resources.

3.8 Statement of Key Ethically Based Factors and Principles

Based on the diverse ethical issues raised above, we have tried to crystallise out a compact set of key messages or conclusions that should be taken into account when formulating or reviewing a regulatory framework for management of radioactive wastes. It is our conviction that these points are too little emphasised in the continuing debate on radioactive waste disposal and also that few if any other industries could frame and defend an equivalent ethical position.

3.8.1 Intragenerational Equity

- Radiation protection criteria for operating disposal facilities should be no less and no more protective than for other nuclear facilities;
- Harmonised nuclear and non-nuclear regulation, based on relative risks and hazards, should be an objective of a rational national programme;
- ALARA should be considered by repository implementors and regulators, but it is not normally a major issue in disposal where expected exposures are very small;
- Host communities should be fairly compensated for accepting a repository;
- Implementors and regulators should allow the public open access to information;
- Implementors and regulators should allow the public to participate in decision processes — without however abdicating scientific and technical responsibility;
- There are no valid ethical arguments against transfer of wastes between willing countries which have the necessary technology for safe management.

3.8.2 Intergenerational Equity

- Current generations should not knowingly undertake actions that will expose future generations to risks that would today be deemed unacceptable;
- Freedom of choice for future generations should not be unnecessarily restricted by the actions of current generations. This means that the issue of retrievability of disposed wastes should be addressed;
- Maximising future choice and minimising future burdens are to some extent contradictory objectives and a proper compromise is needed; this should be well documented and agreed by all interested parties;
- Ethical debates on radioactive waste disposal should also involve considerations of environmental and resource-conservation benefits to future generations from the use of nuclear energy by current generations;
- Funding for future nuclear liabilities should be assured; segregated funds for this purpose are a common mechanism. The further the planning or implementation of repositories advances, the more credible are the estimated costs of final disposal.

3.8.3 Conclusions from Other Ethical Principles

- Nuclear power — with a solution to the problem of permanent disposal — is a practice which satisfies the requirements of sustainability;
- The public should be made aware that waste repositories, provided that they are not interfered with, cannot fail in a catastrophic manner which endangers either the environment or future society in a fundamental manner;
- The concentrate and confine philosophy in geological disposal makes quantification of the costs much easier than in many other technologies. Financing schemes for radioactive waste disposal should ensure that the waste producer pays.

Chapter 4

Reversibility and Retrievability

The concept of deep geological disposal was developed in order **to permanently remove** radioactive wastes from the human environment. Repositories with multiple passive barriers (engineered and geological) are designed to ensure that the wastes remain isolated from the human environment and inaccessible to man for the very long times needed to allow for the natural decay of their radioactivity. The very foundation of the concept is that wastes deep underground will be contained until they present no significant hazard. Retrievability was therefore not a significant issue during concept development.

Retrieval of wastes for safety reasons was reckoned by disposal experts to be a scenario of such low probability that little effort was devoted to its study. Retrieval for other reasons, such as recovery of usable raw materials (fissile isotopes, precious metals etc.) was treated under the heading of deliberate human intrusion. The philosophy that was commonly followed was that no measures should be taken to ease such retrieval and that any future society deliberately embarking on this course is itself responsible for any risks arising. The responsibility of today's society is to maximise the safety and security of future generations whilst imposing minimum future burdens. The security angle is particularly relevant for repositories which contain fissile materials either in the form of conditioned plutonium from weapons dismantling or spent fuel with its residual fissile content. Repositories of this type, if maintained in a state where retrievability is easy, require constant application of safeguards measures.

As pointed out in the previous chapter, however, in recent years there has been an increasingly active debate on what exactly are the prime responsibilities towards future generations by the current one. Do we want to minimise the burdens or maximise the choices of options — or can both aims be fulfilled at the same time? Can fully passive (and safe) systems provide a sufficient level of practicability of retrievability? Should implementors plan for enhanced future accessibility in order to offer wider choices or should they emphasise passive safety systems that may make access more difficult, but will thereby minimise future burdens?

This debate is linked directly to practical, technical matters, such as the design of the facility, the operating procedures and the institutional programmes (including monitoring) throughout the lifetime of a repository (see Chapters 10 and 11 on institutional control and monitoring). But there are also philosophical issues involved, in addition to these purely technical issues. Most importantly, there is a growing recognition that many societies are uncomfortable with the concept of perceived irretrievable disposal; bitter lessons from the past have too often revealed that technical or societal developments have not always progressed as expected, and that yesterday's solutions to problems can have environmental impacts that are found unacceptable today.

Thus we have a perceived potential conflict. Technologists are dedicated to avoiding any compromise of safety that might be caused by introduction of intrusive, post-closure monitoring or of engineering measures to facilitate retrieval that might be counter-productive. Society at large has less confidence in technology and a stronger desire to keep options open. The public, moreover, is also not convinced of the experts' view that current designs already provide a significant level of safety combined with enough scope for reasonably straightforward retrievability.

Discussions in dedicated working groups such as the IAEA group on Principles and Criteria (IAEA, 1997b) or in special fora (e.g. the EU Concerted Action on Retrievability; see Grupa et al., 2000) or in ad-hoc groups (e.g. NEA, 2001d) have tackled the key issues directly. For retrievability, the questions are:

- How easy does retrieval have to be in the different stages of repository development ("staging" varies from programme to programme: Box 3 describes typical stages in a repository development programme)?
- What is the rationale for requiring a given level of retrievability at any specific phase?
- What technical retrieval measures and methods are feasible?
- Should specific features facilitating retrievability be introduced into the repository design?
- How do such measures impact on other aspects of system performance and on other issues (such as nuclear safeguards)?
- Do funding arrangements need to be set in place for provision of longer "open" periods for a repository, for retrieval operations, and for subsequent management of retrieved wastes?

In fact, the intense, relatively short-lived debate on retrievability has been beneficial in developing sensible waste management policy and presenting it in a positive light. The debate has made most groups think more closely about the way disposal will actually be managed over the many decades of operation of a deep waste repository. It was initially seen by many as a new and different conceptual basis for waste management. There was an early, negative or defensive response from the technical community. However, closer examination of disposal concepts led to the realisation

Box 3: Typical Stages in a Repository Development Programme

Surface exploration: is used both to distinguish between different potential sites and as part of the detailed characterisation of a candidate repository site. The objectives of this work are to provide a comprehensive understanding of the nature and properties of the geological and surface environments which can be used to support safety assessment and basic repository system design. Surface exploration would continue until confidence in the potential of a candidate site was sufficient to move to the stage of underground exploration.

Access construction and underground exploration: Construction of an underground repository starts with the excavation of the access shaft or adits and the preliminary layout of access galleries. Reconnaissance and investigation work within this stage will supplement site characterisation data acquired during the surface exploration. Design of the repository system will be optimised at this stage of the pre-operational phase. Significant perturbation of the natural system would be expected to occur as underground excavation commences.

Construction of the repository: This is the main pre-operational stage, in which excavation of waste emplacement galleries, disposal vaults, shafts or boreholes is undertaken. This may be carried out as part of a single construction campaign, or be a progressive programme, with waste being emplaced in some regions of the repository while others are still under construction. In some repository development programmes, it is envisaged that construction of the main repository might be preceded by a pilot stage, in which demonstrations of technology can be made to enhance confidence in the concept. This approach might be linked to the construction and commissioning of parts of a repository specifically designed for intense, long-term monitoring, throughout the open life of the repository.

Emplacement of waste and near-field engineered barriers: This stage begins with the commissioning of the repository system, followed by a lengthy period (typically decades) of operation. The main activity is the emplacement of the waste packages within their immediate engineered barriers. There are different options for the time at which these various barriers may be put in place, depending on waste and rock characteristics. Any national approaches to requirements for waste package retrievability may have a significant influence on the options chosen. However, it is important that these latter considerations do not prejudice the design basis of the near-field engineered barrier system. For example, delay in emplacing barriers immediately adjacent to waste packages might, in many concepts, lead to less than optimum performance in the post-closure period.

Disposal tunnel/vault backfilling: Timing of and procedures for backfilling and sealing of sections of a repository where disposal has been completed will depend again on national decisions on retrievability and on constraints dictated by the properties of the host rock. It could occur concurrently with continued construction or disposal activities in other sections of the repository. This may allow filled sections of the repository to be directly backfilled to isolate individual emplacement tunnels or vaults.

Repository backfilling and sealing: Repository backfilling and sealing constitutes the final stage in closing a repository. All access ways including shafts will be backfilled and sealed to isolate the disposal vaults and cells. The decision to close the repository will depend on a number of factors including technical considerations, societal choices and the safety consequences of keeping the repository open. The decision of how and when to proceed with repository closure will be a matter of national policy.

Post-closure (institutional/non-institutional): The post-closure phase will begin when the repository access ways have been backfilled and sealed. Some programmes may choose to begin the post-closure phase with a period of active institutional control. With or without such a period, monitoring and surveillance could be maintained for as long as society considers it beneficial, although (as noted above) it is a principle of geological disposal that assurance of safety does not require post-closure monitoring.

that retrievability is always an option: a fact of which the public is not well aware. A careful, stepwise operational strategy can be devised that has options for pausing, taking stock and reversing actions at every stage, without necessarily having to compromise safety in the long-term. When analysed, it was seen that this was the direction that most repository operations would have taken in any case, simply to make progress through the hurdles of acceptance and permissions. The result is that retrievability is coming to be regarded by many implementors as a manageable issue that can be embraced as integral to their methodologies.

The following discussion looks at the questions identified above, the positions that have been taken by various interests, and how the recent debate has dealt with them.

4.1 Rationale for Retrievability

It is possible to advance **technical** arguments for retaining a post-closure retrievability capability in a repository. The most obvious argument is that, despite all the safety features in the system, the repository might not perform to the predicted standards, with the result that radionuclides are released in unacceptable concentrations. This scenario pre-supposes that monitoring methods have been established to detect any leakage and that an evaluation of the safety has led to the conclusion that the release levels justify remedial action by retrieving the wastes. This scenario is regarded as incredible by many designers and analysts of repositories; however, monitoring to enhance public confidence in safety is accepted as necessary in many programmes (see Chapter 10).

A period during which the wastes in their final configuration can be observed, monitored and if necessary retrieved with relative ease has, in fact, been a feature of regulations in some national programmes (USNRC, 1983b). The feasible timescales, however, were judged to be only some decades; whilst this is long for human activities, it covers only a negligible portion of the relevant containment timescales for a geological repository. Technical arguments have also been made concerning recovery of valuable constituents, including fissile materials, after a long period of cooling has made the wastes more amenable to handling and treatment. Clearly, pre-supposition of a need to recover fissile materials contradicts the requirement to eliminate these materials (make them "practically irrecoverable") or else to maintain nuclear safeguards permanently.

If recovery of waste for any purpose is explicitly foreseen, however, monitored storage on or below the earth's surface may well be a more obvious approach than geological disposal. Further quasi-technical reasons advanced for maintaining retrievability concern the potential of new, as-yet-undiscovered technologies. A new method of eliminating radioactive wastes might emerge, a hitherto unforeseen application for some constituents of the waste could become important. The counter-arguments to such ideas are more philosophical than technical and are addressed below.

The **ethical** arguments related to final disposal have been increasingly debated in recent years (see Chapter 3). The starting position was clear and is documented in various international consensus documents (IAEA, 1995b, NEA, 1995a). Wastes should be managed by the current generations (who enjoy the benefits of the corresponding nuclear applications) in such a way that the burden on future generations is minimised. Deep disposal in a passive repository system from which retrieval is not foreseen was the proposed answer. Initiated largely by ethical discussions in Sweden (KASAM, 1988), an alternative view emerged in the 80s. This view is that we have an even higher responsibility to future generations — namely to give them the widest possible choice of societal options. By making retrieval from a repository more straightforward, the range of future options is extended. The burden imposed by extra future measures is claimed by some to be outweighed by the benefits of wider choice.

This broad moral argument may, in fact, be a rationalisation of **societal** arguments based on the subjective feelings of a large segment of the population that is still sceptical that geological disposal will fulfil the high safety standards set. The timescales for disposal are too long to be comprehensible; technologies have failed unexpectedly in the past; neither the risks, nor the costs, nor the time pressures associated with prolonged storage are unbearably high. Given these perceptions, a societal strategy postponing final decisions is tempting and understandable. Responsible technologists must respond to societal wishes, therefore disposal plans will inevitably have to address the issue of retrievability. In practice, this is being done in many national disposal programmes. Retrievability is demonstrated by engineering studies or even practical trials. Normally the difficulty of retrieval increases with time throughout the stepwise process of emplacement, backfilling, sealing and long-term monitoring.

A final, very pragmatic reason for retrieval options being built into disposal concepts is that corresponding **legal or regulatory** requirements are in force. These can reflect a judgement on technical reliability (e.g. US requirements for an initial 50-year retrieval period, USNRC, 1983a) or on ethical priorities (see for example the Netherlands law making retrievability compulsory or the positions of the German, Dutch and UK Governments, given in NEA, 1994).

4.2 Measures to Enhance Retrievability

Use of the word "enhance" in the title of this section alludes to the fact that geological disposal, per se, is always retrievable in principle. It is important to recognise that this fact is not generally known to the public. At question is the length to which the implementor goes to ease retrievability. The effort involved in retrieving disposed wastes is directly affected by the strategy and the technical concepts chosen. For example, easiest retrieval is achieved by delaying disposal and maintaining surface storage, whilst options like sub-seabed disposal make retrieval more difficult, or practically almost impossible. The choice of host rock is important. Stable self-supporting crystalline rocks are less problematic with respect to retrieval than soft clays which creep, or salt formations which are also plastic. A long-lived container with radiation shielding capability will make retrieval simpler. A soft backfill allowing easy re-excavation will do likewise. Studies have been made on techniques for removing waste containers from clay backfills (e.g. Kalbantner and Sjöblom, 2000). Specific examples from different national programmes are given in IAEA (2000).

It is also possible to conceive engineering designs that aim at easy retrievability by automated excavation tools. This approach could affect the repository layout, the sealing techniques as well as the backfill, buffer and waste package. Long-lived overpacks, packages with pre-mounted handling attachments, tunnel liners dimensioned to remain intact for long periods, are all examples of engineering approaches to easing retrieval. Further possible measures include high-resolution, near-field monitoring (although current technologies would be invasive and will inevitably be unreliable over even a few years of operation), and comprehensive data recording and archiving (see Chapters 10 and 11).

In summary, geological disposal is always retrievable in principle — but numerous specific measures can be implemented in order to enable stored or disposed wastes to be retrieved with increased ease. However, any decision on retrievability must also consider the impact on other aspects of the disposal system.

4.3 Potential Impacts of Retrievability

Enhancing retrievability can obviously have significant impacts on the design of a repository and on the operational procedures. The impact of retrievability measures on the long-term, post-closure safety of a repository is a major issue since steps taken to keep the wastes accessible for retrieval may, in fact, negatively affect the isolation capacity of the repository system. In addition to obvious risks due to postponement of backfills, seals, etc. there are other technical disadvantages which can arise. A repository kept open for decades to ease possible retrieval is subject to geochemical changes due to the oxidising environment, rock mechanical effects, increasing hydrological perturbations etc. These effects can degrade the long-term safety performance and/or make it harder to model this performance with an adequate level of confidence.

Also affected by retrievability measures is the operational safety of a repository. Obvious examples are hazards associated with flooding, gas build-up, mining safety, etc. The potential risk to health and safety at a filled repository will certainly be reduced by bringing the system into its final, sealed configuration as soon as possible, even if this increases the difficulty of subsequent waste retrieval.

To many minds the additional radiological hazards resulting from maintaining a repository in a state allowing easy retrievability are less of a concern than the increased safeguards risks associated with the potential misuse of radioactive — and particularly fissile — materials. Specific studies (Peterson, 1998), have illustrated that clandestine retrieval even from fully closed and sealed repositories may be feasible. Easily retrievable spent fuel, especially as it cools with age, could become an increasingly attractive target for rogue governments or for dedicated terrorist organisations.

More mundane drawbacks of retrievability proposals are in the financial areas. Explicit engineering measures to enhance retrievability or to postpone final sealing and decommissioning, inevitably give rise to extra costs. The procedures for putting the repository into its final passive safety configuration (backfilling and sealing) may be more complex and more expensive if they have to be carried out in a repository with all wastes already in place. These higher costs may, to some extent, be offset by the fact that they are deferred to a later time. For the implementor, a key management and financial issue will be the point at which responsibility for the repository passes from them, to society or the state. A funding mechanism must, therefore, be established to ensure that any delayed costs can be met.

More important than providing for the costs of delayed completion of repository closure measures, moreover, is covering the cost of potential retrieval if this is reckoned to be a real future option. In fact, this funding discussion can be taken further. If the reason for retrieval is inadequate performance of the repository or development of a new improved disposal method, then funding for subsequent actions should logically also be secured. Supporters of geological disposal who seek to enhance acceptance of the concept by offering full retrievability options at all future times should not disregard the potentially far-reaching financial implications of this commitment. The cost and financing implications of implementing a retrieval policy are discussed specifically by McCombie (2000) and Söderberg (2000).

A related aspect to the availability of funds for closure following a long "retrievability period" in which the repository is incomplete, is that of the availability of expertise and, even more fundamentally of the will and the interest to finish the project as planned. It is not inconceivable that social changes, wars or catastrophes over the tens of years in the operational life of a repository could reduce the capacity to close a repository satisfactorily. Funds may be diverted elsewhere, priorities may change or there may simply not be the technical resources available any longer — an issue now recognised more widely in the nuclear sector (NEA, 2001d). This is clearly an ethical issue to do with protecting future generations whilst we have the will and the capability to solve the problem. It argues strongly against adopting any engineering retrieval measures in which the eventual safety of the whole disposal system relies on final closure actions.

On the other hand, it is clear that public acceptance of repository projects does indeed depend upon both the actual and the perceived degree of retrievability. The nuclear community has had very limited success in communicating the basic concept of geological disposal. The laudable, ethical objectives of implementing facilities which do not require active monitoring to provide safety and from which retrieval need never be foreseen are often misunderstood. A common public perception is that monitoring will not be carried out and that retrievability is impossible. These misconceptions must be countered by an open discussion which includes recognition that public doubts must be taken seriously, describes the procedural and engineering measures which can be taken to enhance retrievability and lays out the advantages and disadvantages associated with these measures.

Finally, perhaps the most likely feature of a disposal programme to be amended because of pressure to make the waste retrievable and the procedures reversible is the timescale leading to ultimate closure of the repository. Given that there is little technical urgency to implement disposal and given that the build-up of public confidence is a slow process, there is an understandable tendency in national disposal programmes towards extended schedules by implementing a series of discrete phases.

4.4 Positions on Retrievability Taken in Selected Countries

As mentioned earlier, the earliest formal position taken on retrievability was in the USA where a 50-year period of retrievability was required in regulations as a guarantee that recovery options were possible should some unforeseen problem occur during the operational period of a geological repository. As the debate on retrievability intensified over the last 10 or more years, the implementing organisations of some national programmes voluntarily built into their concepts easier retrievability.

In Sweden, SKB amended its strategy to include a 25-year demonstration disposal phase and specific studies were performed to provide evidence that wastes could be retrieved after this period if this choice were made. For a Swedish or Finnish repository, with long-lived containers embedded in soft bentonite clay within a stable hard crystalline host rock, this is a relatively straightforward matter. Other countries also addressed the technical feasibility of retrieving emplaced wastes, e.g. UK Nirex studied the removal of soft grouts from around ILW containers in a deep repository (Brenn and McCall, 1997).

In Switzerland, Nagra, in response to public opinion in the wake of the 1995 referendum on the Wellenberg L/ILW repository, introduced design and operational features to allow easier monitoring and retrieval of wastes for decades or even centuries. Also, the USDOE OCRWM organisation has altered the reference design of the proposed Yucca Mountain repository to make direct observation of individual waste packages and easy retrievability feasible for at least three hundred years (DOE, 1998).

At a regulatory level, the tendency was still to warn against the possible negative safety effects of easy retrievability rather than to require that retrieval be possible. For example the Swiss regulations (HSK & KSA, 1993) state that, whilst retrievability is not forbidden, any measures intended to ease retrievability may not have a detrimental effect on long-term safety. The regulatory situation changed when the authorities in the Netherlands forbade any geological disposal (of any hazardous waste) which was not shown to be retrievable (Selling, 2000). This tendency of authorities to respond to public pressure requiring retrievable disposal has grown stronger with time. France requires retrievability now (CNE, 1998). The latest report of Government experts in Switzerland proposes test disposal facilities for L/ILW and HLW (EKRA, 2000). It accepts that wastes are inherently retrievable and recommends that disposal caverns in the main repository are sealed as soon as is feasible with separate test and pilot caverns being used for test and demonstration purposes. In Germany, the fact that salt as a host rock creeps to completely seal the waste canisters is being used as a negative argument, since this complicates retrieval (Brenneke, 2000).

The current situation worldwide concerning retrievability is that virtually all countries will expect to be assured that retrieval is feasible if required. The question of financial responsibility for any retrieval operations has not been cleared up, although it would be expected that any incremental costs associated with specific provisions for retrievability during the operational stage would be a small component of the overall cost of disposal (IAEA, 2000). The tendency is that the disposal organisation continues to be responsible for as long as it exists — in some cases it may also make financial provisions in case retrieval is made necessary by malfunctioning of the repository. Ultimately, the long timescales of relevance imply that responsibility must pass to the state.

4.5 Conclusions

It will be decades before deep geological repositories come into operation, they will operate for many more decades and might be sealed only after a protracted monitoring phase. Accordingly, there is little operational pressure to finalise retrievability concepts. Indeed, as noted in the introduction, the issue of retrievability appears to be of considerably less significance now than when it first arose, largely because it is now appreciated (at least within the industry) that a careful, stepwise disposal process can always be reversible. Nevertheless, disposal systems are being actively planned and designed, so that retrievability features do need to be discussed now. More importantly, the whole issue of retrievability is irrevocably linked to the question of public confidence in the safety of geological repositories — and this fundamental issue is directly linked to the ethical and environmental questions concerning continued use of nuclear technologies. Retrievability as part of a wider concept "reversibility" is currently being discussed in various groups working on the concept of staged repository development (e.g. NRC, 2003).

Opponents of deep disposal would prefer to leave wastes indefinitely in monitored surface or underground stores. Proponents argue that this is not a sustainable solution, that it is a higher risk option and that one should proceed in a stepwise fashion towards final disposal. In the current climate of opinion, it may be possible to move forward only if the question of retrievability is tackled head on. Any disposal project submitted for approval should discuss the balance drawn between minimising future burdens and maximising future options; explicit features which ease or complicate retrieval should be pointed out; the cost as well as the cost–benefit of any retrieval option should be addressed. A strategy which allows confidence in the safety of disposal to be built up gradually throughout a series of phased steps has the greatest chance of acceptance — even when these steps involve decreasing levels of retrievability.

The following conclusions are formulated in a manner intended to focus discussion on the issue of retrievability of wastes from deep geological repositories.

- Public opinion is such that disposal projects should directly address the issue of retrievability/reversibility through all phases of repository development. Retrieval is always possible in principle. Engineering methods to allow retrievability are available, even though they become more complex and expensive as the step-wise closure of the repository progresses and with increasing time after closure of the repository. Implementors should be prepared to take measures to assure the public of these facts on the basis of specific studies on retrieval concepts and techniques.

- Measures to ease retrievability may have a negative impact on long-term, passive safety and security. It is a responsibility of repository designers and analysts to make this clear to the public and to decision makers. It is also their responsibility to ensure that any impact is limited to justifiable levels. The future generation that eventually decides to complete and decommission a repository must be comfortable with the decision that they (not us) will be making. Whatever we decide now, there is no compulsion whatsoever for eventual operators and regulators of a repository to adopt our philosophy or respond as we do to present-day drivers. Thus, there will be considerable opportunity for changes in approach to decision-making before a repository has reached the end of its operational life. What does remain our responsibility is to ensure that future operators can complete the task safely, perhaps with their own changes, and certainly in their own time, rather than leaving them with an incompletely designed facility that is not intrinsically safe at all times, both operational and post-closure.
- The most obvious method of retaining maximum retrievability is by extended, or "indefinite" surface storage. This approach does not, however, represent a solution to waste management. It postpones burdens and responsibilities into the future in a manner incompatible with a sustainable development ethic. Storage is, nevertheless, an important step in the waste management process. A step-wise closure process for a repository, including retrievable storage periods on the surface and/or underground at the chosen site, can maintain the sustainable concept of passive long-term safety that minimises future burdens, whilst still providing for a lengthy transition period and an appropriate level of reversibility/retrievability. This gives sufficient time for societal decision-making on the path towards final closure of the repository.
- For HLW without a significant content of fissile materials, retrievability arguments are related mainly to the confidence of different groups in the long-term safety performance of the repository. For fissile materials, the prime arguments for and against retrievability concern resource conservation and weapons safeguards. Retrievability and assured nuclear safeguards are clearly completely incompatible objectives. However, the public desire to have reversibility as such — without specifying the reason or giving any justification — needs to be acknowledged.
- The social and technical process for decision-making for closing a deep geological repository (and for reacting to low probability scenarios involving potential remediation measures, up to and including retrieval) has never been completely defined. However, it is envisaged that an institutional programme will address: (i) the type of activities to be performed at the different development phases (*in situ* monitoring, complementary and confirmatory research programmes, periodic re-evaluation of safety, etc.) (ii) the criteria and decision-making process (licensing, etc.) to react on these activities, (iii) the options (including retrieval) available at each decision-point.

- Directly tackling the issue of retrievability can help ensure that repositories are developed in a step-wise or phased procedure which allows time for organisations and individuals involved to build up a high level of trust, based on open communication and on demonstrably high-quality technical work. However, stipulation of retrievability (or reversibility) is not a logical component of national regulations, particularly as there will always be an intrinsic possibility in a step-wise programme to reverse each step anyway. As noted above, there may be societal demands that put such requirements on the implementor, via laws or government policy statements, but these are a step removed from regulations. However, if national policy requires some element of retrievability, then the regulations must account for the impacts that this might have on the safety of the system, and should consequently require the implementor to demonstrate impacts on repository performance at each stage of the operational and post-closure life of the system.

Chapter 5

Timescales in Repository Evolution

In all discussions on geological disposal, a fundamental issue concerns the timescales to be considered in safety assessments and regulations. Various timescales are of direct relevance for concepts involving deep disposal of long-lived radioactive wastes. A common feature is that the times to be considered are far longer than humans are used to thinking about when judging technical systems. An example of the considerations involved is evident in the Japanese AEC Guidelines for a major study of HLW disposal completed in 2000 (AEC, 1997), which state:

> "... it was considered to be appropriate ... to evaluate radioactivity releases ... without specific limits on the timescale of safety assessment. Timescales should be determined from the standpoint of long-term changes in the human environment, the long-term stability of the geological disposal system and the potential hazard associated with HLW."

What exactly are these timescales? The direct time-dependent issues affecting regulations for long-term safety are:

- changes in radioactivity, or more relevantly radiotoxicity, of wastes with time;
- time dependence of the evolution of engineered and geological safety barriers;
- time dependence of the evolution of the biosphere, including human society;
- times at which peak doses or risks are calculated to occur, when analysing typical repository systems;
- times at which the integrated probability, even of rare disruptive events, as described in Chapter 8, becomes significant.

The credibility of long-term safety analyses is also affected by more general considerations such as:

- the objective technical reliability of current scientific methodologies for quantitatively analysing the future evolution of the safety system;
- the more subjective confidence of scientific, public and political bodies in technical analyses extending into the far future.

In the latter context, most people are seriously concerned about the safety of future generations no further than their grandchildren: less than a 100-year time frame. A recent study of public opinion in Japan, UK and Switzerland (Duncan, 2001) showed that 75–80% of people who were questioned thought only this far forward when considering the future welfare of themselves and their family, and more than 90% only looked as far as 500 years into the future. The latter time horizon was also cited by more than 90% of people when considering a wider social perspective: the future welfare of their township. In Switzerland and the UK 80–90% of Swiss and UK respondents asked about their timescale of concern for the global environment indicated that this stopped at 1000 years.

Given the short forward timescales with which people are concerned it is not surprising to hear that the public worries are about whether a repository next door will irradiate people each time they drive past it; will it in the near future leak and poison their water; will it contaminate their community's land? This suggests that we should apply much effort to reassuring people about what we (quite perversely, in the public view) would call "short-term" safety. The most recent Eurobarometer survey of attitudes to radioactive waste in 15 EU countries (INRA, 2002) showed that, overall, about 58% of those who had a view on living near a deep repository were most concerned with transport of waste to the site, leaks during operations or reduced property values. The remaining 42% were concerned about environmental impacts over the next hundreds or thousands of years (which the poll, interestingly, in the light of the comment above, did call "long-term" effects).

A need to focus more on the immediate future, when interacting with the public, is reinforced when we consider the frequent poor experience of prediction over years or decades. Experience of scientific prediction is that, within a person's lifetime, things very often work out differently to how one was told they would. People are understandably sceptical about the experts' claims to predict the integrity of passive, man-made, engineered systems for more than a few decades into the future, unless constant maintenance is assured. Going back to the study by Duncan (2001), it was found that the majority of those questioned believed that waste could only be contained in a repository for 100 years. This suggests a rethink of the conventional way that we present safety assessments could be a valuable contribution to enhancing understanding.

Of course, for scientific analyses and — critically — for regulatory purposes, the immense timescales of hundreds of thousands of years must be considered by the experts. The discussion of these should not, however, completely swamp the more immediate issues raised above. There are also other relatively short timescale issues of relevance for developing regulations for repositories. These include the times for which simple retrievability might be required in the pre-closure stage, the times for which monitoring and/or institutional control measures may be required in the post-closure phase, etc. However, this chapter — rather paradoxically, given the caveats just raised — now concentrates on the most problematic issue, namely regulations for long-term safety.

5.1 Relevant Timescales for Analyses

In terms of the long-term behaviour of a repository, the most quantifiable parameters are the half-lives of the radionuclides in the wastes and their consequent heat output. Consideration of typical decay curves for spent fuel or HLW (see Chapter 2) has led to two important time periods being commonly identified. Decay of the shorter lived, but highly active radionuclides like Cs-137 and Sr-90 for 10 half-lives (i.e. ~300 years) gives a factor of about 1000 reduction in activity and in heat generation. A 1000-year period of substantially complete containment[10], as mentioned in various regulations, thus ensures that the shorter-lived radionuclides have essentially decayed and the main thermal pulse affecting the repository and surrounding host rock is past.

At longer times, as discussed in Chapter 2, the decay of radioactivity is more gradual. A common point in time identified later is when the activity or toxicity of the original uranium ore used to produce the fuel and waste is reached. The timescales used for this differ depending upon the normalisation used (e.g. per unit mass, volume etc.) and upon values used for activity or radiotoxicity. As a round number, a figure of one, to a few hundred thousand years is a justifiable value. Although return of the radioactivity to the original ore levels gives a valuable figure of merit, it is recognised that this need not correspond to a negligible risk level, since the wastes may be more concentrated than the original ores, and are certainly at a different location.

Much more subjective judgement is needed in quantifying the time periods for which credible analyses of the behaviour of engineered or geological barriers can be performed. Waste forms, such as the ceramic matrix of spent fuel or borosilicate glasses, are predicted to resist dissolution for tens of thousands, even millions of years, in typical repository environments. Reference container lifetimes of thousands of years (steel construction) up to hundreds of thousands or even a million years (copper) have been used in formal safety assessments. Natural materials, such as bentonite, which is proposed as a buffer material, are expected to be stable for millions of years. Estimates for all these engineered barrier lifetimes are based on laboratory experiments, *in situ* experiments and, perhaps most convincingly, observations on natural analogue systems (see Box 6 and Miller et al., 2001).

The stability of the geological environment and repository host-rock formations is a key issue, which is obviously highly dependent on the regions being considered. Internationally, the stability of some of the environments in which deep disposal is contemplated have been little affected by tectonic processes over times up to tens, even hundreds of millions of years, but it is difficult to interpret the geological record over such long periods sufficiently well to allow precise predictions of future evolution. In practice, the period of stability must be judged for each of the

[10]Substantially complete containment does not preclude the early failure of a small fraction of waste containers; doses resulting from such isolated failures must be within regulatory limits.

important geological processes or events that could initiate or influence releases of radionuclides from the repository. These include faulting, uplift, erosion, glaciation, volcanism, etc. As discussed in Chapter 8, there is broad international acceptance that the future likelihood of occurrence and nature of impacts of tectonic events and processes can be scoped using data from the past 10^5–10^6 years. The precision with which they can be predicted depends, of course, on the process concerned, the period of analysis and the geographical region considered.

Most problematic are the time spans over which the behaviour of the large-scale and fine-scale hydrogeological systems influencing groundwater movement and potential transport of radionuclides can be predicted with sufficient reliability. The nature of these systems sensitively affects the performance of a geological repository. They are difficult to characterise with sufficiently fine spatial resolution even for present day conditions. The timescales over which the hydrogeological systems may change (due, for example, to precipitation or dissolution of minerals or to climatic changes) is even more difficult to assess.

The most rapidly changing part of the overall repository system is obviously the biosphere. Major climatic changes, including glaciation, can occur over tens of thousands of years and affect most regions of the world (see Box 4). Lesser, but significant climate and biosphere changes, can occur over hundreds of years — and may become even faster due to anthropogenic emission of gases leading to a greenhouse effect. As discussed in Chapter 2, human society evolves fastest, and it is impossible to predict confidently the living and eating habits of societies for even tens of years into the future.

Box 4: Climate Change

Geological history is characterised by infrequent periods in which Earth is gripped by glaciation. These epochs occur irregularly, separated by hundreds of millions of years. The reasons why and when they occur are not fully understood, but are related to factors such as the global distribution of land and ocean.

Some of Earth's glacial epochs last for millions of years and are characterised by cycles of warming and cooling, leading to periods of advance and retreat of ice. We are in a warm period within such a glacial epoch at present. The current epoch, and its cycle of warming and cooling has occupied approximately the last two million years and is likely to continue for at least some hundreds of thousands of years into the future. The cycles within glacial epochs are caused by the superposition of a number of "forcing factors", including changes in Earth's orbital characteristics around the sun. The current, Quaternary glaciation is characterised by intervals of extensive ice cover in the northern hemisphere at periods of glacial maxima, and warmer, interglacial periods when ice cover is much reduced. The figure below shows the estimated volumes of ice in the northern hemisphere over the last 200,000 years and the expected volumes over the next 150,000 years.

The most recent glacial maximum was 18,000 years ago; the previous about 135,000 years ago. We are currently in an interglacial period; the previous interglacial temperature maximum was 120,000 years ago. The figure suggests that another major period of ice cover will begin affecting northern hemisphere landmasses (and cause a global fall in sea levels of about 150 m) from about 50,000 years into the future, peaking at about 100,000 years. Recent research has begun to provide more detail, indicating many sudden cold events and warm (interstadial) periods superimposed upon the overall trend. In the historical period, there have been less pronounced but well-documented minor fluctuations in climate, such as the so-called "little ice age", most marked during the 16th and 17th centuries when (for example) the River Thames froze over almost every winter.

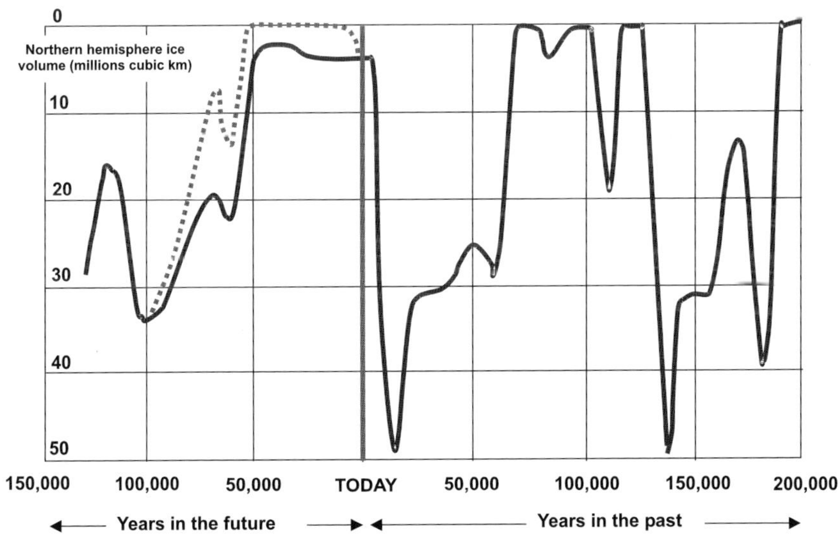

Long-term changes in global temperature, indicated by estimated past and future volumes of ice in the northern hemisphere. Warm (interglacial) periods, such as the present day, have low to zero ice volumes. The dashed line shows one possible future where marked global warming melts the Greenland ice sheet. Note, however, that the long-term trend remains the same (after Berger et al., 1998).

"We're not taking this very seriously, are we?"

5.2 Calculated Timescales for Releases from Repositories

Analyses of the isolation capability of typical deep repository systems commonly calculate peak releases to the biosphere hundreds of thousands or even millions of years in the future. The assumed release scenarios directly influence the calculated times to peak release. Making the assumption that some waste containers fail early, as a result of imperfect quality assurance, can lead to predicted releases after times of around 1000 years. For some systems, e.g. disposal in salt, the normal scenario is zero release for all conceivable times in the future and the safety assessments often focussed on inadvertent intrusion scenarios (see Chapter 9).

The confidence that can be placed in such performance assessment calculations is the subject of much debate (e.g. NEA, 1999a). Much of the scepticism often expressed by the public concerning scientists' ability to quantify future repository performance stems from the fact that the results are often presented or interpreted as precise "predictions". The realistic picture is that they are only broad estimates, that they normally quantify the bounding behaviour for any given evolution scenario and that, if this is done for a range of scenarios representing the principal processes and events that might affect repository evolution, a credible safety analysis is possible.

5.3 The Problem of Compliance with Regulatory Criteria

Problems can arise when attempting to transfer all of the above scientific considerations into the regulatory area. Typically, a regulatory process involves comparing the expected future behaviour of the disposal system with some form of compliance criteria. In particular, in countries with a strong adversarial legal system, the issue becomes the defensibility of compliance arguments in a court of law. The point has been made that regulation out into the far future requires either soft criteria to be compared with the results of rigorous assessments, or else hard criteria, with subjective judgement allowed in assessing compliance. In practice, as discussed in Chapter 6, the trend in most countries is to aim at a combination of hard quantitative criteria together with soft qualitative measures of performance. Bringing these together obviously requires judgement.

One approach that has been proposed is to subdivide the future into various time slices and to use different approaches to judging acceptability, and differently weighted performance measures, in each time frame. Examples from national regulations given in the table in Section 5.4 illustrate possibilities such as:

- Calculating doses using rigorous analysis with the present day biosphere for 1000 years;
- Calculating doses using a reference biosphere out to 10 ka;
- Calculating doses using a reference biosphere out to peak dose;
- Using comparisons with natural radionuclide fluxes in the geosphere over a period of reasonable predictable geological stability, e.g. 100 ka to a million years or more (see Chapter 8).

The fact is that there can be no complete mathematical proof that a disposal system will perform as calculated far into the future. This has led to the use of terms like "reasonable assurance" or "reasonable expectation" in regulations related to licensing repositories. In addition, as discussed in Chapter 6, discussions on acceptability should be based on a *range* of information and quantitative estimates of performance measures.

Current ICRP views on time slices and "potential exposures" are that assessments should be mainly qualitative and decisions should be based on reasonable assurance; calculated doses should not be used as absolute compliance targets but as indicators of potential hazards.

5.4 Current Situation Internationally

The most intensive regulatory discussions on timescales have taken place in the USA. The Environmental Protection Agency (EPA) has used 10,000 years as a compliance period in regulations for various environmental areas, including in the regulation 40 CFR 191 (EPA, 1993), under which the first custom-built repository for long-lived wastes (WIPP) was licensed. A working group of the National Academies in advising the US Government on regulation of Yucca Mountain considered that there was no scientific reason for the 10,000 years and recommended analysing longer timescales (NRC, 1995). In its recent draft legislation for a Yucca Mountain standard (EPA, 2001c: 40 CFR 197), EPA has retained the 10,000-year limit for compliance calculations but specified that results calculated out to the time of peak release should also become part of the public record of the licensing procedure. Because these debates are on-going and because of the past tendency of many countries to follow US licensing approaches, at least in the nuclear reactor area, more details of the USA situation are given in Appendix 2.

An extreme counterpoint to the USA position is exemplified by the Swiss regulations R21 (HSK & KSA, 1993). Here the basic requirement is that the given dose or risk limits may be exceeded *at no future time*, i.e. there is no time cut-off to the analyses. However, presentations of PA results by Nagra emphasise the increasing uncertainty with time (Nagra, 1993b).

Table 5.1 gives a summary of how different countries have handled the issue of timescales in their national regulations for repositories (derived from NEA, 1997).

Table 5.1

Timescales in national waste disposal regulations

Canada	Time frame for quantitative compliance 10,000 years; requirement that longer periods be addressed qualitatively to ensure that no sudden increase in risk would occur
Finland	In an assessment period that is adequately predictable with respect to assessments of human exposure but that shall be extended to at least several thousand of years" dose constraint from expected evolution; beyond, quantities of radionuclides migrating to be below specified limits (derived from natural backgrounds)
France	Stability of geological barrier to be demonstrated for a period of at least 10,000 years; calculations of dose for normal evolution extend to 100,000 years; thereafter the situation is "hypothetical"
Germany	No time frame officially specified; recommendation of the RSK for dose calculations to 10,000 years and use of other safety indicators thereafter
Sweden	High demands on quantitative risk calculations using present day biosphere for first thousand years; thereafter risk calculated to not less than 10,000 years but does not have to exceed 1 Ma. Particular attention required to period of next glacial cycle (100 ka)
Switzerland	Doses and risks shall "at no time" exceed specified values
UK	The official guidelines specify a risk target for the post-closure period which is of undefined duration. The advisory body, NRPB, has proposed different approaches for different time periods (NRPB, 1992)
USA	40 CFR 191 (EPA, 1993) specifies dose limits for 1000 years, cumulative release limits for 10,000 years, groundwater permissible concentrations for 1000 years; 10 CFR 60 (USNRC, 1983a) specifies "substantially complete containment" for 300–1000 years, water travel times of at least 1000 years; 40 CFR 197 (EPA, 2001c) requires compliance demonstration for 10,000 years, presentation of results to peak dose or risk

It is interesting to note that the period of 10,000 years that enters into regulations in various countries, is not obviously derivable from any of the time considerations at the beginning of this chapter. In practice, the period corresponds to an estimate of the time since the last glaciation and the assumed time to the next. It is thus a biosphere milestone rather than a key period for deep geological changes — although the changes in local and regional hydrology, world sea levels etc. associated with ice ages could obviously have far-reaching effects on some disposal systems.

5.5 Conclusions

When setting timescales for which the performance of a deep geological repository must be analysed and for which compliance with regulatory criteria must be demonstrated, it is important to differentiate between choices based on policy and choices based on science. There are no real scientific grounds for specifying any specific time cut off for either safety assessments or regulations, beyond which there is no requirement to consider the fate of the repository. In particular, any cut-off imposed whilst calculated releases are increasing has no credibility.

An approach commonly used is to calculate releases, doses or risks out to peak consequences — but to use different approaches to judging acceptability in different time frames. A reference biosphere is in all cases necessary when dose or risk calculations extend even 1000 years into the future (see Chapter 6). At far future times (>10 ka), dose or risk calculations using a reference biosphere are still a useful way of putting calculated releases into perspective in a stylised calculation. The calculated doses may then be more appropriately compared with less stringent limits than the typical limits at shorter times (e.g. 10^{-6} risk or 0.1 mSv/a dose). For example, comparisons may be with natural radiation doses or radionuclide fluxes. For each radionuclide, it is also informative to show its ultimate "fate" as it is dispersed in Earth's crust or into the oceans. For extremely long-lived radionuclides in the natural radioactive decay series, this "fate" is likely to be to remain within the repository, like a natural ore body, until it is eroded away and they are naturally dispersed. Later, in Section 6.6, we suggest an approach to setting containment objectives and associated standards, that addresses the timescales issue directly.

It is vital that the increasing uncertainties involved in any estimates at distant times in the future are presented properly, and accounted for sensibly, when using the results to make regulatory decisions. There are, however, legitimate arguments that legal compliance decisions may be more difficult when considerations must include times in the very far future for which it can be argued that our knowledge of the system state is increasingly uncertain. Especially in adversarial legalistic licensing systems, this could lead to problems.

Chapter 6

Performance Measures and Appropriate Standards

This chapter discusses the different styles of *performance measure* that can be used to judge the behaviour of a repository and which should be considered for inclusion in regulations for long-lived waste disposal. Some of these performance measures are directly related to radiological safety criteria specifying limits on radiation exposures of present or future individuals (i.e. dose and risk measures). Other measures of the performance of the repository system are less directly associated with radiological safety but may be more meaningful or understandable when applied to far future times (e.g. comparisons with natural geological concentrations or fluxes of radionuclides). The trend today in new regulations is towards incorporating both types of measure.

Current thinking on safety assessment for geological repositories would classify **all** types of performance measure as *safety indicators*. This includes the most commonly used measures of dose and risk. In the past, dose and risk have more commonly been regarded as hard safety criteria, with other measures thought of as less quantitative (and, perhaps, less valuable or rigorous) indicators. However, quantitative assessment of dose or risk requires specifying details of human lifestyles that may change drastically on the long timescales involved in repository safety analyses. Accordingly, it is acknowledged that precise numbers calculated must not be over-interpreted. With the increased understanding of the uncertainties surrounding predictions over long time periods, each of the measures discussed in this chapter is now being regarded as part of a spectrum of useful indicators that all have a part to play in judging repository safety and, specifically, in implementing a regulatory system.

6.1 Radiation Doses and Risks

As indicated in Chapter 1, international recommendations concerned with the safe disposal of radioactive waste are based on the well-established principles of radiological protection. These principles derive from recommendations issued from time to time by the ICRP and encapsulated in the IAEA Basic Safety Standards (IAEA, 1996a). Although these would seem the obvious basis for national regulatory criteria, in recent years the ICRP claims that national policies have, in fact, not commonly been derived from their advice (ICRP, 1998b). This may be due to the fact that the ICRP recommendations tend to be developed in a rather small and somewhat isolated "radiation protection community", and have often been rather theoretical and difficult to translate into practicable regulations. Some of the problems associated with communicating issues associated with radiological protection and the different approaches of experts were touched upon by the OECD Nuclear Energy Agency Committee on Radiation Protection and Public Health (NEA, 2000d):

> ...the public, politicians and decision makers often are less able to understand the different approaches. Public concern over radiation exposure does not seem to be related to the level of dose incurred, as shown by the low concern over medical exposures as compared with the public outcry over very low exposures from the clearance of radioactive waste.
>
> Of relevance to the system of radiation protection is the increasing social desire/need to understand decisions made by governments, regulatory bodies and industry, and to participate more actively in decision-making processes involving environmental and health issues. To address this need, industry, governments, and regulatory bodies are becoming increasingly transparent in terms of their operation.
>
> Radiation protection is no exception to this trend. Scientific rationale that was once sufficient to explain radiation protection theory and practice is no longer adequate. The need to address and communicate theory, practice and the decision-making process to a wider audience has led to numerous debates and led the radiation protection community to revisit the framework of the system of radiation protection. The very fundamentals of the system of radiation protection continue to be questioned in a healthy fashion, and many aspects have been identified which could better serve stakeholders given some additional thought in the light of modern societal needs.

At present, the fundamental system of radiological protection is contained in ICRP Publication 60 (ICRP, 1991a). This system divides activities into practices and interventions[11], with waste disposal being classed as a practice. ICRP 60 states

[11] A practice is any deliberate activity that introduces additional sources of exposure, or modifies exposure pathways. An intervention is a remedial action to reduce or avert exposure, or the likelihood of exposure, to sources which are not part of a controlled practice.

three principles that should be applied to practices, which in brief can be summarised as:

- JUSTIFICATION: practices should not be introduced unless they produce sufficient benefit to offset the detriment, in terms of radiation exposure;
- INDIVIDUAL DOSE LIMITS: individual exposures from all relevant practices should be subject to dose limits, or to some control of risk, in the case of potential exposures;
- OPTIMISATION: the magnitudes of individual doses, or risks in the case of potential exposures, and the numbers of people exposed and the likelihoods of exposure, should be kept as low as reasonably achievable (ALARA), taking into consideration social and economic factors. (This principle is difficult to apply in practice and is not widely applied in waste disposal: see previous and later discussions.)

Justification is assumed not to be required in radioactive waste disposal; the practice is part of the larger practice of nuclear power production so no special case needs to be made for disposal alone. It can be seen that the principal underlying quantitative criterion for regulatory decision-making is *individual dose*. As discussed later, the definition and application of this measure for radioactive waste disposal is complicated.

The appropriate dose to apply is the *effective dose*, which is the sum of the equivalent doses for all parts of the body exposed to all types of radiation for the range of radionuclides that might be considered in assessing a release pathway from a repository (see Box 5). In the subsequent discussion, the word "dose", used on its own, will thus mean the effective dose to an individual.

If it is not certain that a given situation will lead to radiation exposure, then the probability of its doing so should also be considered, along with the dose that would result. The combination of the dose and probability of receiving this dose can be defined as *risk*. In radiation protection, a more relevant and understandable measure of risk is defined by also factoring in the probability that any dose received will lead to a health effect (or, less euphemistically, to a death). This last factor is the dose-to-risk conversion factor, discussed in more detail later.

Thus, we have:

Health risk, R, from an event	=	Probability, P, of the event occurring	×	Dose that would result from the event	×	Probability of death per unit dose

The simple definition of risk as the probability multiplied by the consequent dose implies that frequent events leading to low doses present equivalent risks to rare events with high dose consequences. Although the numerical risks may be the same, society reacts differently to these two types of risk exposure, and this has led to much discussion.

Box 5: Radiation Protection Quantities

Activity (A): the amount of a radionuclide can be expressed in terms of activity, which is the average number of spontaneous nuclear transformations (decays or disintegrations) taking place per unit time. The unit used is the **becquerel** (Bq), which is equivalent to one transformation (disintegration) per second.

Absorbed dose (D): the fundamental dosimetric quantity is the energy imparted by incident ionising radiation per unit mass. The unit of **absorbed** dose is the **gray** (Gy), which is equivalent to one joule per kg.

Equivalent dose (H): the absorbed dose delivered by a specific type of radiation over a tissue or organ, multiplied by a weighting factor for the type of radiation which reflects the effectiveness of different types of radiation in inducing health effects (neutron and alpha radiation having a much higher weighting factor than beta and gamma radiation). The unit used is the **sievert** (Sv).

Effective dose (E): the sum of the tissue equivalent doses, each multiplied by the appropriate tissue weighting factor which reflects the different sensitivities of different organs and tissues to the induction of stochastic effects of radiation. The highest weighting factors include those for the gonads, bone marrow, lung and stomach, and the lowest those for the skin and bone surface. The unit used is also the **sievert** (Sv).

Dose limits: currently recommended dose limits from the ICRP (1991a) and the IAEA Basic Safety Standards* can be summarised in outline (there are special limits set for example, for young workers, for certain tissues, and for certain circumstances that are not detailed here) as:

Exposure	**Annual Effective Dose Limit (mSv)**
Occupational: averaged over five consecutive years	20
Occupational: in any single year	50
Members of the public	1

*IAEA, 1996a, from which the information in this box is taken.

6.1.1 A Note on the Broader Context of Risk

An extensive literature exists on the broader subject of risks in modern society. In this book, we address only the technical issues associated with setting of standards. Much more has been written, however, on issues concerning the public perception of, and acceptance of risks. Concerning the former, the main point of relevance in the present context is that nuclear risks are perceived by the public in general as being much higher than other, more familiar risks. A single, striking illustration of this can be seen in Fig. 6.1 which, for an American public, illustrates the discrepancies between actual (statistically quantified) risks and the corresponding perception of different groups.

The acceptance of risks by the public is obviously strongly coloured by their familiarity with the risky activity in question. It is also influenced very strongly by other factors such as the degree to which the risk is imposed by others rather than voluntarily accepted by the risk takers. There are many obvious examples of this: for example, smokers who show great concern over minute quantities of food additives.

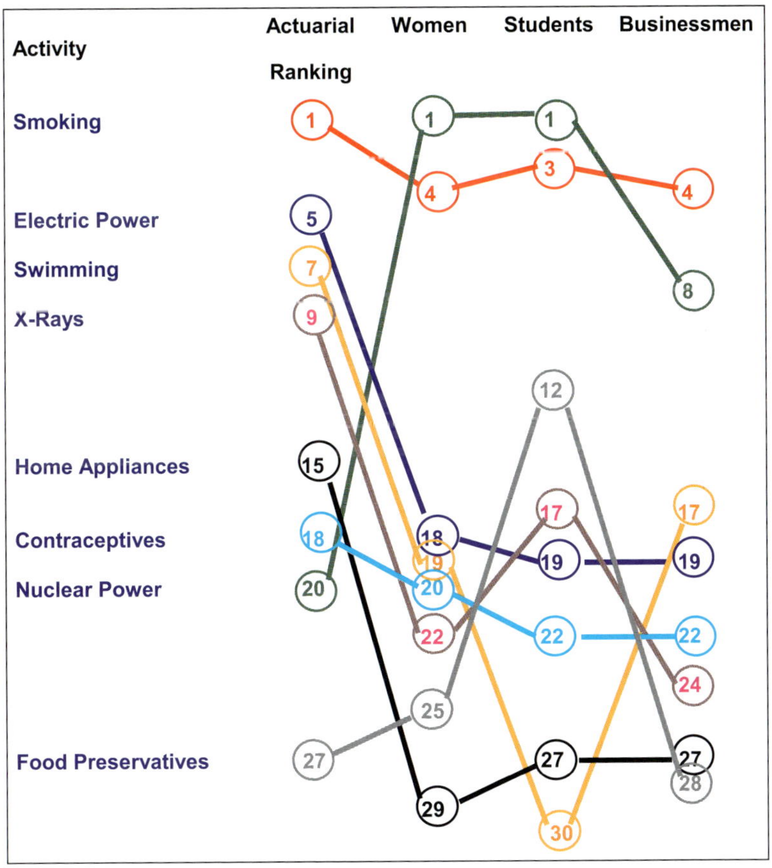

Fig. 6.1. The disparity between the risks of various activities as perceived by different groups of people and the actual risks (as quantified actuarially), is graphically illustrated in this survey from the USA. A selection of some of the 30 risks studied is shown. (Data from Slovic et al., 1985.)

More relevant to the standards issues dealt with in this book is the different levels of radiation risks that are regulated as being acceptable for voluntary workers in the nuclear industry and for involuntarily exposed members of the public. The task of setting levels of risk which are to be regarded as acceptable by workers or by the public is controversial and challenging. It is, however, a decision of a political or societal nature and will not be directly addressed in this text. A good overview of risk issues is provided by NRC (1996).

6.1.2 Dose Limits and Constraints

People will normally be subject to a number of sources of radiation exposure. Thus it is useful to define an individual dose **constraint** for exposures resulting from a single source, such as a waste repository. A dose constraint should only be a fraction of the overall recommended radiation dose **limit**. The ICRP's recommended dose limit for members of the public (for all exposures to man-made, non-natural sources of radiation: i.e. doses that can be *controlled*) is 1 mSv/a, and this figure has been incorporated into the IAEA Basic Safety Standards (IAEA, 1996a). ICRP (1998b) notes that the application of dose limits to waste disposal has intrinsic difficulties. They thus recommend that emphasis is placed on **dose constraints**, and that a process of **constrained optimisation** will obviate the direct use of the public exposure dose limits in the control of radioactive waste disposal.

"Constrained optimisation" is a difficult piece of jargon that does little to make radiological protection principles transparent. It is meant to take into account the fact that optimising according to ALARA (see Section 3.5.2) could in principle lead to an unequitable distribution of benefits and detriments. For example, in waste disposal those people in the future receiving detriments are not the same as current population receiving benefits. To prevent an unfair shift of detriment, a dose constraint is applied. This is a value (lower than the dose limit) that may not be exceeded for any person, even if a net optimisation could thereby be achieved.

The ALARA principle has a further important condition built into it, namely that the optimisation should be done "taking account of social and economic factors", which acknowledges that people's values, and their perceptions of the benefits and the cost implications of dose reduction measures, need to be considered too. When considering how tolerable risks are to society, there can also be a lower bound below which it is not worth optimising further. This is at the level where social and economic perceptions and responses combine to make the risks involved with very low doses "negligible" or "below concern". Below this bound, society would not normally consider it worth expending additional resources to reduce risks further. The spectrum of doses or risks between the upper and lower bounds is sometimes described as the "risk tolerability" region, or the "ALARA region"

[12]ALARP: as low as reasonably practicable: in practice, this means almost the same as ALARA: as low as reasonably achievable, but, by changing the term, emphasises the ALARA suffix "taking social and economic factors into account" (i.e. practicable).

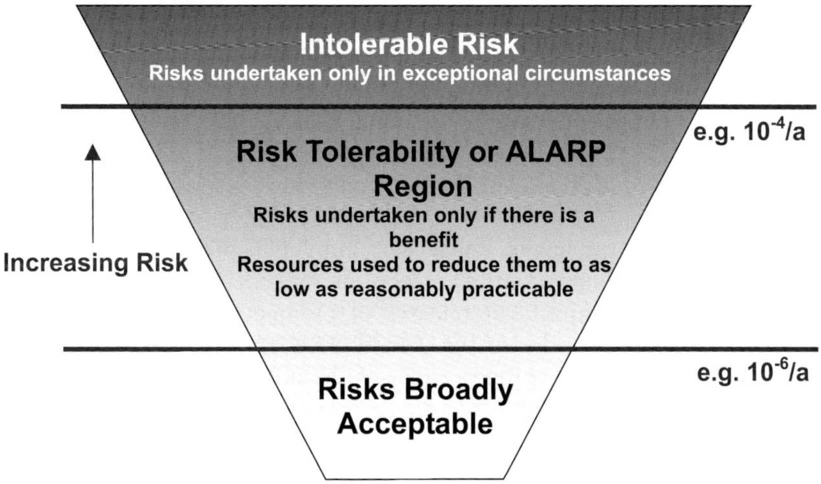

Fig. 6.2. The "risk tolerability" framework, expressed as annual risks to individual members of the public. The actual values of risk considered to be tolerable or intolerable to society vary from one activity, industry or practice to another. In the upper part of the ALARP region it is clearly worth expending resources to reduce risks. This becomes less worthwhile in the lower region, as the benefits of risk reduction compared to the costs involved diminish. In the "acceptable" region, it is considered not worthwhile using resources to reduce risks any further.

(or ALARP[12] region: see Fig. 6.2). As will be seen later, the dose and risk constraints proposed or used for geological repositories are often very close to the "below concern level", making the ALARA region in which constrained optimisation could be undertaken extremely limited. This contrasts with the way in which society reacts to other (non-nuclear) risks, where the "risk tolerability" region is much broader because higher levels of risk are considered tolerable.

The commonly applied dose constraint for a waste repository in a number of national regulations is 0.1 mSv/a, although ICRP suggests that a value of 0.3 mSv/a would be appropriate. One rationale that has been given for using a lower constraint for repositories than for other nuclear facilities is that any doses that do result from a closed repository will be far in the future, and to populations other than present generations, which are benefiting from the nuclear practices giving rise to the wastes. In assessing the safety of a waste repository, a common view is that individual dose constraints should be applied to the value of the average dose in the maximally exposed group (i.e. in the group that is representative of those individuals expected to receive the highest dose). Others prefer applying it to the maximally exposed individual, even though this can lead to unrealistic and impracticably strict regulations. The definition and use of **exposure groups** is discussed later.

An obvious issue in setting dose constraints for a HLW repository is that the site may contain more than one repository. There may, for example, be a spent fuel and/or a long-lived ILW repository attached, close by. ICRP 77 (ICRP, 1998a) says that the source to which the dose constraint is applied "*should usually be... the whole repository*", which could be taken to mean all related disposals at the site. This decision may become a matter of national policy.

6.1.3 Collective Dose and Negligible Incremental Doses

Apart from looking at the health impacts on individual people, under the current ICRP optimisation recommendations, decision makers should also have some measure of the overall impacts of releases of radioactivity. The quantity called **collective dose** was introduced into the general area of radiological protection for this purpose. Collective dose is the product of the number of people exposed to a source and their average radiation dose (expressed in man.Sv). The calculated collective dose is a measure of the total detriment or harm to the population for which the collective dose is calculated.

Clearly, the principle of optimisation of radiological protection, in which the number of people exposed, as well as their individual exposures, needs to be kept ALARA, suggests that collective doses should be kept low. This use, in the context of deep geological disposal, where collective doses (unless specifically restricted in space and time) may be accrued over many generations and extremely long periods of time, is controversial and is discussed further, in Section 6.3.

Collective doses, particularly for geological or ocean disposal of wastes, sometimes express very low incremental doses (i.e. far below natural background: see Box 6 on natural radioactivity) to large numbers of people. One of the reasons for the reduced relevance of collective doses (see Section 6.3) is that, at the low doses expected from radioactive waste repositories, uncertainties in the dose-effect relationships are large, which make the uncertainties of the impacts of collective doses so large as to become more or less meaningless. There is much debate at present concerning whether there is some level of dose that can be regarded as negligible, or so low that it can be ignored.

Part of the debate centres upon the unknown effects of low doses of radiation. All direct observations of harmful effects are at much higher doses than those regulated and there is no firm scientific information on the potential health effects, such as increased cancer rates, due to low doses of ionising radiation. Indeed the concept of hormesis (Mossman, 2001; Calabrese and Baldwin, 2003) asserts that low-level exposures may actually be beneficial, or even essential, for living organisms. This discussion is particularly prevalent in the USA. Part of the debate results from the sensitivity of the public to the issue. For example, the USNRC was forced to withdraw suggestions for defining doses that were "below regulatory concern" (BRC) because this was interpreted by many as neglecting the concerns of the public. A further aspect of the debate on low dose levels is the application of so-called "trivial" levels of dose, at the current clearance/exemption level, to waste disposal regulations, which, for example, is reflected in current Japanese regulations for LLW.

Box 6: Comparisons with Nature and History: Analogues in Natural and Archaeological Materials

Comparisons with nature are helpful in several ways. First, the backbone of radiological protection discussed in this chapter is effectively drawn from setting exposure limits that are fractions of the natural radiation "background" to which we are all exposed. Second, some regulatory standards make comparisons of releases of radioactivity from a repository with natural concentrations of radionuclides in environmental systems (e.g. in various "compartments" or regions of the biosphere), aiming at ensuring that they are only a small fraction of these levels (see Section 6.5.2). At a more basic level, the overall safety concept for the longest-lived and most radioactive wastes such as spent fuel generally aims at ensuring containment until the wastes are broadly similar in activity to natural uranium ores (see Chapter 2). In addition, more direct indications of how radioactive wastes might behave over very long periods of time, immensely greater than those that can be used in experiments, can be found in natural geological systems such as ore deposits and in archaeological materials that have been buried for thousands of years.

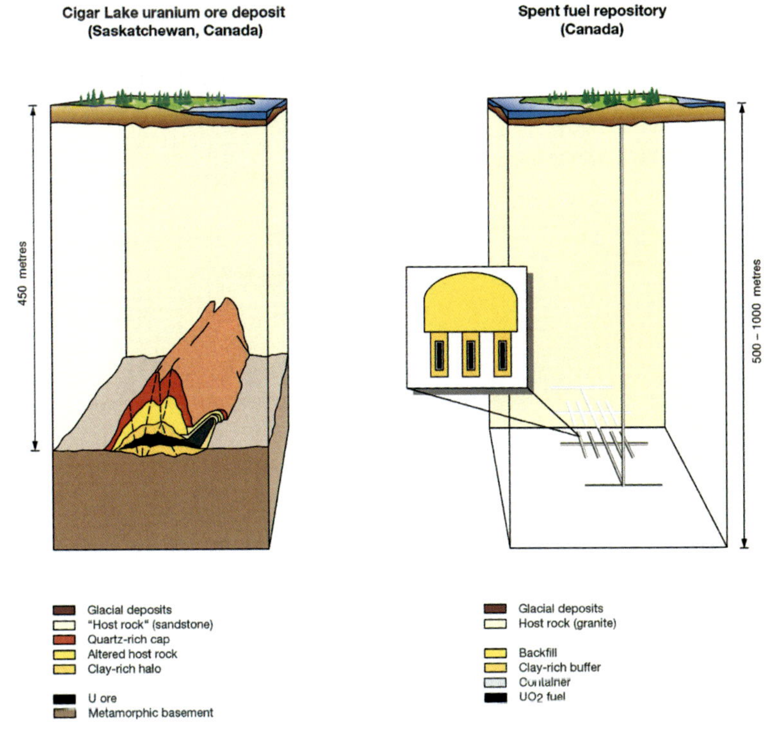

(Goodwin *et al.*, 1989)

Comparisons with nature can provide compelling evidence of the isolation capabilities of stable, deep geological environments. For example, the Cigar Lake uranium ore body (see Cramer & Smellie, 1994) shows many similarities with concepts for spent fuel disposal. An extremely high-grade uranium oxide ore (analogous to spent fuel) is surrounded by clay (analogous to clay buffers around spent fuel containers) at a comparable depth to those being considered for repositories. The ore is about 1.3 billion years old (ten thousand times longer than the containment period typically aimed at for spent fuel), yet remains contained within the clay "halo". The clay has limited the flux of water through the ore, despite active groundwater movement in the overlying sandstones, and the ore has an extremely low solubility. In fact, there is no direct radiochemical evidence of the existence of the ore body at the surface. The stability of the geological environment over a very long period has maintained conditions at depth that are favourable to containment of the radionuclides in the ore.

Archaeological materials (especially metals, concretes and glasses) also provide useful evidence that helps surmount the "time barrier" encountered in assembling experimental data on, for example, corrosion rates. Artefacts that have been buried in chemically stable environments, such as are considered for the engineered barriers in a repository, are often well preserved. An appropriate archaeological "find" can provide direct confirmation of, for example, corrosion rates of massive iron containers for HLW, over exactly similar timescales to those assumed for container longevity (typically hundreds to thousands of years). The seven-tonne hoard of Roman nails from a legionary fortress at Inchtuthil, Scotland, discovered in the 1960s, (above) provided just this kind of evidence. The nails and other iron artefacts were buried in a pit to keep them from the local tribes when the Romans withdrew from the area. Much of the material in the centre of the pit, although corroded on the surface, was otherwise intact (the outer regions of the mass of iron removing free oxygen from the groundwaters, keeping the core region chemically reducing) and allowed estimates of corrosion rates over a period of about 2000 years (see Miller et al., 2001).

The ICRP has always stated, and continues to believe (although there is a substantial body of evidence and opinion of the contrary view), that there is not a lower dose threshold below which there is no risk of stochastic health effects. The US National Council on Radiation Protection and Measurements adopts a similar view, noting that there is no conclusive evidence on which to reject the linear, no-threshold (LNT) dose–response assumption, although observing that more data are needed (NCRP, 2001). They also point out that, at the same time, the probability of effects of very low doses (e.g. natural background) is so small that it may never be possible to prove or disprove the validity of the LNT assumption.

The LNT position is adopted by ICRP as a prudent default option, but it has been pointed out that it can lead to massive diversion of resources into areas which do not represent a significant public health risk (Rockwell, 1997; Kellerer & Nekolla, 2000). ICRP believes that it is not possible at present to move away from an LNT position without a futile debate which would damage the credibility of the radiological protection profession (Clarke, 1999). In fact, as discussed later, ICRP is considering new concepts for controlling radiation doses in which the existence (or not) of a threshold would become largely irrelevant. At the moment, their policy for waste management is based on limiting the risk from stochastic effects, not on eliminating them entirely, which means that risk limits must be defined on the basis of broader public acceptability of all types of risks.

6.1.4 Potential Exposure and Risk

All of the problematic issues outlined above arose originally in considering low-level radiation exposures to current generations. When we consider waste disposal, two further key points emerge: doses may, in fact, never occur and, if they do, they will be far into the future. This raises the issues of timescales and probabilities.

Before the integrated system of radiological protection was brought together in ICRP 60, recommendations with respect to solid radioactive waste disposal were contained within ICRP 46 (ICRP, 1985), which itself was based on the underlying system of dose limitation recommended in ICRP 26 (ICRP, 1977). ICRP 46 dealt specifically with the issue of long timescales, and with the uncertainties involved in identifying if, and how, radiation exposures might occur in the distant future. In doing this, it introduced the concept of **potential exposure** to cover the possibility, but not certainty, that some scenarios for the future evolution of a repository may expose people to radiation. It recommended that dose limitation be applied to the "normal evolution" of a repository: that is, to the expected degradation of the safety barriers over time. However, ICRP 46 introduced the concept of **risk limitation** to protect future generations in respect of probabilistic events (and environmental changes) and potential exposures. In terms of the long-term behaviour of a deep waste repository, risk is clearly linked to the probability of an event or change occurring that could initiate a release of radioactivity into the environment (thus resulting in *potential exposures*). Thus, for probabilistic events or changes, the risk to an individual is the product of the probability that such an event/change will occur and the probability that it will cause a serious health effect.

For many years, the ICRP position was that the 1 mSv/a dose limit implied a risk of the order of 10^{-5}/a, which meant that a dose constraint of 0.1 mSv/a would be equivalent to a risk constraint of 10^{-6}/a (this is only about twice the risk of being struck by lightning — see Chapter 13). The conversion from dose to health risk is based primarily on analyses of people who were exposed to doses much higher than the low values considered here (in particular, A-bomb survivors from Japan). The extrapolation to lower doses depends sensitively on the dose reconstruction studies done to assess the kinds and levels of radiation experienced by these people. Re-evaluations of these factors have led to the dose to risk conversion factor being increased. Various authorities cite slightly different values for the "dose-to-risk" conversion factor for fatal cancers. UNSCEAR (1993) give a figure of 4×10^{-5}/mSv, a figure that is also implied by USEPA cancer incidence risk factors. ICRP 60 gives a figure of 5×10^{-5}/mSv. ICRP also gives factors for the risk of non-fatal cancers (1×10^{-5}/mSv), and serious hereditary genetic effects (1.3×10^{-5}/mSv), which gives an overall cancer and hereditary effect risk factor of 7.3×10^{-5}/mSv. The same overall cancer incidence risk factor (6×10^{-5}/mSv) is used by the USEPA. A commonly used "rule of thumb" for the risk of "death or serious health effects" that derives from these figures is that a dose of about 0.015 mSv is equivalent to a risk of 10^{-6}. A risk of 10^{-6}/a is significant because it is used in a number of national regulations for radioactive waste disposal.

There is thus a direct connection between dose and health risk, which is implicit in the dose limit recommendations of ICRP. Dose limits are chosen to correspond to the upper boundary of a range of *risks* that are considered to be acceptable. These risks include a variety of human activities. Thus, dose limits were derived from acceptable risks at the outset.

ICRP 46 and 60 are complementary documents that, together, form the current radiological safety basis for radioactive waste management. However, recent developments need to be considered, particularly when setting up a new regulatory framework that would be based on ICRP recommendations, and these are discussed further in Section 6.2.

6.1.5 Exposure Groups and Reference Biospheres

The estimation of radiation doses to people necessarily involves some level of modelling of radionuclide transfer pathways in the biosphere. However, there are large uncertainties involved in biosphere modelling, especially with respect to the future behaviour of human populations, which cannot be predicted with any confidence for periods of more than a few decades. To address these irreducible uncertainties, the international BIOMOVS and BIOMASS projects, managed by the Swedish Radiation Protection Institute, SSI and the IAEA respectively, have developed the concept of **reference biospheres**. This concept also includes suggestions for the definition of **exposure groups** (e.g. BIOMASS, 1999b).

For present day releases of radioactivity, from existing practices (e.g. a nuclear power plant), radiation doses are calculated to **critical groups**. Instead of calculating the distribution of doses over the entire exposed population, only doses to the most exposed groups of the population are calculated. The aim of this long-established approach is to reduce the analytical effort required, whilst still ensuring that adequate radiological protection is maintained.

For present day releases from any nuclear installation, the critical group is defined as follows. First of all, the habits of the population in the region into which the radionuclides are (or will be) released, or in the region where they might accumulate, are investigated. Based on this information, the group that is expected to receive the highest doses as a result of its way of life is defined as critical group. Doses (or risks) are calculated for this group. By ensuring that doses to the critical group are below a certain value, every individual in the population is protected. For nuclear facilities that are currently operating and emitting radiation, the critical group might be the community living closest to the facility, the community closest downwind or downstream, or a community with special dietary habits (e.g. consumers of seaweeds, which tend to concentrate radionuclides from sea water).

For present day releases the habit surveys show which groups are the "critical ones". Releases from a properly implemented radioactive waste repository can occur only in the far future. Human habits will certainly have changed significantly, and habit surveys cannot be carried out. Therefore, proof that the groups defined for releases far in the future are the critical groups is not possible. Therefore, it is now preferred to use the term **exposure groups**. The basic question is how to define the hypothetical individuals within the exposure group who are to be protected from future releases. In this context, three types of uncertainty or variability are relevant: variability of time of releases, spatial variability of releases and variability in human habits.

Since any releases from a repository are likely to occur over protracted periods compared to a human lifetime, ICRP (2000a) suggests that it is not necessary to estimate doses to different age groups as the average over a lifetime is adequately represented by annual dose or risk to an adult.

Considerations of variability in *time* are often simplified by considering only the time of either peak releases or of peak concentrations in the environment, or pre-defined timescales for which different assessment procedures have been prescribed. Given the long timescales before releases occur, changes in the environment and climate must be anticipated.

A release of radioactivity will also cause a *spatial* distribution of concentrations. In general, the aim is to protect the persons at the location(s) at which the highest concentrations occur. For releases in terrestrial environments, one or more such regions of high concentrations can generally be easily identified, and they will probably be limited in size. Releases into a marine environment are likely to impact (although at a very low level) larger areas than terrestrial releases, in particular, for releases of long duration.

Variability in *human habits* exists at several spatial scales; at a global scale, between countries, at a continental or even smaller scale, between populations and communities within countries and, finally, within communities. These variabilities are partially caused by historical, cultural and economic factors and partially by environmental factors, such as climate, topography and the presence and distribution of terrestrial, fresh water and marine environments. Further variability is caused by differences in habits as a function of age. All these can cause large variabilities in the distribution of doses that might be received. For the assessment of radioactive waste disposal the following differences in habits may be most relevant:

- **Between different populations, communities or groups:** for example, present and past communities over a range of topographies and climates representative both for today and for future environmental changes (e.g. fishing communities in coastal areas, agricultural communities in mountainous regions) — note that the potentially relevant habits have to be consistent with the environment in which the highest radionuclide concentrations are predicted to occur;
- **Within communities:** for example, people eating no fish, people eating no meat, people eating neither.

As noted above, regulations may be framed such as to protect either **maximally exposed individuals** or **average members of a group**. The ICRP critical group definitions (for references and further discussion see BIOMOVS, 1996) thus imply that the emphasis is on *average members of a highly (maximally) exposed group*. Another approach (used by the USEPA — see Appendix 2) is to define a "reasonably maximally exposed individual" (RMEI). An extreme position is to define an actual maximum exposed individual, making the most pessimistic assumptions possible about his lifestyle. Differences in opinion exist, and a clear position should be established in national regulations. The objectives of implementing safety systems so robustly that no single individual can ever be exposed to more than a trivial dose, however far into the future that individual may live and however unusual his habits, is felt by many to be verging on hubris.

An important point, often misunderstood by critics of dose calculations for far future times, is that performance assessors do not try to predict the exact future course of events at a repository. Rather, by considering a sufficiently wide range of plausible scenarios they try to find any that could lead to unacceptable consequences. With this philosophy in mind, we would make the following suggestions concerning the definition of reference biospheres and exposure groups for calculating doses:

- A **number** of reference biospheres should be defined based on typical environments into which radionuclides may be released, accounting for potential future environmental changes (e.g. in climate). These biospheres can be used to identify typical exposure groups that might exist within them. Such groups should be taken to have similar habits to analogous groups existing in similar environments today
- Our interpretation of international and national radiation protection regulations is that the dose to an average adult member of an exposure group, which is chosen because it is expected to be at the high end of the range of exposure, should be considered. The focus should not be on the most highly exposed individual within an exposure group (e.g. the individual with the highest fish consumption). If the average dose to group members is calculated then, the group should not be so large that the average is heavily affected by many predictions of very low doses
- Because radioactivity will be dispersed throughout the environment, the exposure group should be assumed to exist at the location and at the time where the highest environmental concentrations are predicted to occur (unless it can be demonstrated to be extremely unlikely that anyone could live at such a location)
- For releases in a terrestrial environment, a self-sustaining agricultural community should be considered as one important exposure group. Generally, "self-sustaining" has been defined with respect to food and occupation; however, energy and other materials might need to be considered
- When considering releases into a marine environment, it should be noted that some coastal dwelling communities obtain a large fraction of their food from the sea. These could provide the basis for a first definition of exposure groups, but relevant databases (e.g. on lifestyles and concentration factors in food) may need to be specially compiled
- The effects of changing climates as well as of changing habits could be covered by reviewing habits in other countries and looking at other (more remote) exposure groups.

6.2 Recent Developments in Dose and Risk

Since ICRP 46 and 60 were published there have been several important areas of discussion, and developments with respect to their application to long-lived waste disposal. In 1998 the ICRP issued a new policy document ICRP 77 (ICRP, 1998a) concerned with the disposal of all types of radioactive waste, which clarified some of these issues. However, the ICRP felt that all three of these publications needed to be supplemented, updated and clarified, and two new reports were published in 2000: ICRP 81 (ICRP, 2000a) and ICRP 82 (ICRP, 2000b), which are discussed below. In addition, a debate was initiated in 1999 by the ICRP (Clarke, 1999) concerning a new concept that might be seen as a significant development of the current radiological protection fundamentals. At the time it was first aired, it was termed the **controllable dose** concept. Some of the issues behind this concept have found their way into ICRP 82.

The key issues that these documents address are:

- constrained optimisation over extended time frames and prolonged exposures to (e.g. natural) radiation dose;
- use of dose and risk in the assessment of potential exposures;
- approach to human intrusion (this is discussed in detail in Chapter 9 of this book);
- use of collective dose and the "controllable dose" concept.

6.2.1 Constrained Optimisation Over Extended Time Frames

The overall issue of how to handle time periods in an assessment was discussed in detail in Chapter 5. This section is intended simply to note the current ICRP (and, to some extent, IAEA) views on applicability of dose and risk criteria over extended future time periods.

The basic position of the ICRP is that, given that we should aim to provide the same measure of protection to future generations as we do to present ones, then the same dose and risk criteria should be applied to the associated health detriment. However, ICRP notes that the uncertainties in making calculations of dose and risk in the distant future are such that, for times beyond *several hundred years*, the results should only be regarded as *estimates*. These should be compared with appropriate criteria in a *test* to indicate whether a repository is acceptable given current understanding of the disposal system. *These estimates should not be regarded as predictions of future health detriment*. This reasonable attitude has been adopted throughout the waste management community for many years; it does not, none-the-less, prevent critics of disposal from continually asserting that (unachievable) exact predictions of the far future would be needed to judge disposal safety.

Within this context, assessments should be carried out to compare the consequences of the **natural** evolution of a repository (approaches to human intrusion are considered separately in Chapter 9) with dose or risk constraints. Both ICRP 77 and ICRP 81 recommend an upper value for a whole repository[13] dose

---

[13]As noted previously, an interpretation of ICRP 77 (ICRP, 1998a) could be that all repositories at a single location should be treated as a single or whole repository.

constraint of 0.3 mSv/a, which is now stated to be equivalent to a risk constraint of about 2×10^{-5}/a. ICRP makes it clear that this is not a formal optimisation process. It is a judgemental process taking social and economic factors into account. It should be carried out in a structured manner, iteratively, as the repository development programme proceeds, but it is essentially *qualitative*. The objective is to ensure that reasonable measures have been taken to reduce future doses, to the extent that the required resources are in line with the reductions. Both ICRP, and the latest draft IAEA Safety Requirements for geological disposal (IAEA, 2001b), place frequent emphasis on the word *reasonable*, in terms of measures taken and resources used.

ICRP 81 in fact restates the old ICRP 20 position, that everyone involved in radiation protection should continually ask the question, "*have I done all that I reasonably can do to reduce radiation doses*" — together with the serious caveat from ICRP 21: "*If the next step of reducing detriment can be achieved only with a deployment of resources that is seriously out of line with the consequent reduction, it is not in society's interests to take that step . . .*"

The latest ICRP guidance, ICRP 82 (ICRP, 2000b), covers situations of prolonged exposure to radiation sources. Although not specific to waste disposal (as is ICRP 81), the report does address prolonged exposures to natural sources and residues from practices, so some of the discussion may be relevant to the setting of a dose constraint. The report recommends that in situations where combinations of transitory and prolonged exposures or a build-up over time of prolonged exposures from a source could occur, and where verification of compliance is not feasible, then it may be prudent to restrict the dose constraint to 0.1 mSv/a. Whilst this view is not discussed in the context of releases from a waste repository, and whether this degree of prudence (a factor of three below the suggested ICRP dose constraint) is applicable in that context is debateable.

The approach to constrained optimisation discussed above is generally reflected in current IAEA thinking. IAEA (1997c) recognises that, although the principle of optimisation is valid and appropriate with respect to waste disposal, its application has to be adapted to what is actually achievable in practice. Detailed, quantitative optimisation is not generally achievable, or attempted. In most cases, the choice of disposal options or alternative sites is not determined by explicit optimisation, nor is repository design (in fact there is almost always a degree of redundancy in the barrier system, so that formal optimisation is difficult).

It should be noted that ICRP frequently emphasises that optimisation of radiation doses and exposures should not be confused with rigorous cost–benefit analysis optimisation used in system design. Despite this, even the most recent IAEA review (IAEA, 1997c) does not make this distinction as clear as it could. The current developments in which ICRP suggests moving towards the "controllable dose" concept (see Section 6.3) would also involve de-emphasising the optimisation process, placing it after protection of individuals. In this new model, the first consideration would be to restrict doses to individuals and then make doses ALARA.

6.2.2 Use of Dose and Risk in the Assessment of Potential Exposures

Potential exposures are those which are not certain to occur but which have some (normally low) probability of occurrence. ICRP acknowledges that there are methodological problems and uncertainties in modelling potential exposures for radioactive waste repositories.

When the concept of potential exposures was first applied to deep repositories it was generally assumed that there was some category of behaviour that could be termed "normal evolution", analogous to "normal exposure" in present day practices, in radiological protection parlance. Other exposures resulting from natural environmental events, could be classed as probabilistic. Extensive experience with the use of scenarios to describe possible future evolutionary states of a repository, combined with a growing awareness of the possible impacts of environmental change on releases, even from deep repositories, has led many away from the concept that there is such a thing as a single "normal" evolution of a disposal system.

Essentially, all possible evolutionary scenarios for a repository can have some degree of likelihood attached to them, even if only qualitatively, rather than as numerical probabilities. ICRP now accepts that their previous recommendation in ICRP 46 to treat normal and probabilistic situations separately may not be practicable. Instead, in ICRP 81, they have moved towards two new categories of exposure pathway: **natural** and **human intrusion** (the latter is the subject of Chapter 9).

Although ICRP 77 recognised that the role of potential exposure in risk assessment for long-lived radionuclides is not yet clear, it continued to recommend that the annual individual risk to a critical group for potential exposure (combined with annual individual dose to a critical group for normal exposure) would be adequate for comparing the limiting detriment to future generations with that currently applied to the present generation. ICRP 81 appears to have moved from this position, as it suggests that the constrained optimisation process could calculate *either* risks or **disaggregated dose and likelihood**. Either approach can provide adequate protection. The latter approach is particularly useful for examining, in detail, those scenarios which have low probability, but high potential consequences (see Chapter 8).

The risk-based approach requires a comprehensive evaluation of all relevant exposure situations and their associated probabilities. Many commentators have pointed out the lack of realism in trying to pursue this approach, although it has not prevented some agencies (e.g. in the US programme) from defining sets of scenarios where the individual probabilities sum to unity. Clearly this is fatuous: the addition of another scenario then reduces the probabilities of all the others, which can lead to a theoretically endless process of **risk dilution**, where increasing uncertainty leads to decreasing risks. Risk dilution can also occur through increasing uncertainties in parameter values. For example, in a probabilistic analysis the peak risk calculated at future times can be high if the behaviour of all containers is believed to be so well understood that a precise and narrow band of failure times can be specified. This calculated risk will, however, be lowered if increased uncertainties are assumed in container behaviour, because these lead to a wider spread in failure times.

ICRP 81 has begun to recognise this by preferring the alternative of using disaggregated dose/likelihood calculations, noting that more information may be obtained for decision-making purposes from separate consideration of the probability or likelihood of occurrence of a particular situation giving rise to a dose, and the resulting dose. This is much more in keeping with the concept of viewing constrained optimisation as "*an essentially qualitative*" process. However, ICRP does not completely rule out the use of risk; it continues to suggest either risk or disaggregated dose/likelihood for application to natural exposure pathways.

The "disaggregated" approach would be based on a representative set of scenarios with various qualitative likelihoods. The latter are not precisely quantified probabilities, but the resultant dose plus likelihood presentation provides an appreciation of the radiological consequences of each scenario, balanced against the estimated magnitude of its likelihood. The decision maker would also want to take account of other factors in evaluating scenarios, such as duration and extent of doses. At present, there is not a worked-through example of how such qualitative likelihoods might be presented. Possible approaches to this issue are discussed in Chapter 12, on the treatment of uncertainties.

To summarise, the latest ICRP guidance can be interpreted as recommending a long-term radiological safety assessment calculating disaggregated doses under reasonable, selected test conditions, **as if** they were doses as defined in the normal ICRP dose constraint framework. However, these should be regarded as performance measures or safety indicators (which could be complementary to other indicators: see later in this chapter). Demonstration of "compliance" is not as simple as straightforward comparison of doses with the ICRP constraints, as proof that the system satisfies such criteria cannot be absolute. A decision on acceptability should be based on reasonable assurance rather than absolute demonstration of compliance, and will require a latitude of judgement. At times far into the future, the ICRP dose constraint should be considered as a reference value only, and should provide a basis for such judgements. Critically, ICRP 81 notes that a repository proposal need not necessarily be rejected whenever the calculated impact exceeds the defined constraint. A safety case should be supported and justified by other evidence to determine whether additional measures could reasonably improve protection.

6.3 The Use of Collective Dose and the "Controllable Dose" Concept: Recent Proposals from ICRP

Collective dose has been viewed by some (e.g. NRPB in the UK; NRPB, 1992) as a necessary quantity (along with individual doses and event probabilities) to include within an optimisation and decision-making process. However, summing infinitesimally small doses to essentially infinite populations over geological timescales, then costing the resultant (potentially extremely large) collective dose and arguing that it is worth committing huge resources today to protect the future, is obviously ridiculous.

ICRP began to tackle this issue in ICRP 77 by suggesting that collective doses should be broken down into ranges of individual doses and the periods of time over which they would be received. This approach is part of the overall move towards centring radiological protection on the control of individual doses from "controllable" sources. Proposals for discussion have been published by the chairman of the Main Commission of the ICRP (Clarke, 1999) that aim to integrate and simplify the use of dose across the whole field of radiological protection. These could have some influence on the way in which radioactive waste criteria are set up and utilised. They suggest using a single quantity, called **controllable dose**, which can be applied to all practices, interventions, accidents and high natural exposure situations. Although still the subject of much discussion, some aspects of this concept have found their way into the ICRP 82 report. At the time of writing, ICRP intends to produce a new set of recommendations around 2005, and the current debate on the controllable dose approach will contribute to any revisions to the present system of radiological protection that are recommended at that time (ICRP, 2001).

The revised system being considered is more strongly focussed on "equity-based ethics", and begins with the premise that all individuals have unconditional rights to certain levels of protection (Clarke, 2001). The next consideration then continues to be the control of individual doses. The principle of controllable dose is that if the individual is protected from a single source (e.g. a repository), then that is a sufficient criterion alone for the control of that source. Also, if the risk of harm to the health of the most exposed individual is trivial, then the total risk is trivial — irrespective of how many people are exposed. Clearly, this principle would obviate any need for evaluating collective doses if individual doses were sufficiently low. What is sufficiently low in this context, and how does it fit into the controllable dose concept? Clarke (1999) suggested a simple, unified set of dose constraints, shown in Table 6.1, where they are compared with the diversity of current ICRP criteria that they should replace.

The new system proposes removing the term "dose limit", which is often misunderstood, and using a **Protective Action Level** (around 30 mSv/a) and a series of **Investigation Levels** that would apply (in the repository context) to actions taken to reduce exposures at source or in the environment. The currently applied dose limit for the public would be at the lower end of an Investigation Level range of a few mSv/a that would prompt an investigation to see if anything simple could be done to reduce exposures (effectively, optimisation). Exposures of a fraction of a mSv would be the most that would be allowed to a member of the public from a single source, and there would be no requirement for the 1 mSv/a dose limit to the public. There would still be constraints for each source, and optimisation applied to them.

Table 6.1

Dose level (mSv/a)	Proposed system (Clarke, 1999)	Current criteria
30	Dose should not exceed this "Action Level" and may approach it only if there is benefit to the individual or dose is difficult to reduce or prevent	• Occupational dose limit • Upper radon action level • Relocation intervention level • CT scan
3	There may be a need to reduce or prevent dose, particularly if there is no benefit to the individual	• Lower intervention level for simple countermeasures in an accident (shelters, use of K iodide): 5 mSv/a • Lower radon action level (3 mSv/a) • Average natural background level • Diagnostic X-rays • Dose limit for members of the public (1 mSv/a)
0.3	Maximum dose to an individual who receives no direct benefit from a single source of radiation	• Maximum constraint from a single source • Typical range of variation in natural background (excluding Rn)
0.03	Trivial risk to the individual	• Exemption levels • Clearance levels

The whole scale could be presented in terms of fractions of natural background exposures.

The 1999 discussion paper suggests removing entirely the principle of justification, as it rarely features highly in decision making in any case. Clearer guidance would be needed on optimisation. It is suggested that the principle might be recast as a requirement to **control** the dose to the representative member of the most highly exposed group and ensure that the dose is ALARP. If the most exposed person is protected then everyone else is also sufficiently protected. This approach could be called "Control" and "ALARP". These proposals have received a mixed response from the established community. Following intense debate at a 2003 symposium, (MacLachlan, 2003), Clarke agreed to a less radical approach that introduces the simplified dose constraints but recognises that the current system of separate public and worker doses, collective dose, justification and optimisation should not be abandoned.

ICRP issued a progress report (ICRP, 2001) that develops the approach a little further. It notes that the need for protective action is influenced by the individual

Table 6.2. Bands of concern about individual effective doses in a year (ICRP, 2001)

Band of concern	Description	Level of dose
Band 6	Serious	> 100 × Normal
Band 5	High	> 10 × Normal
Band 4	Normal	1–10 mSv Typical natural background
Band 3	Low	> 0.1 × Normal
Band 2	Trivial	> 0.01 × Normal
Band 1	Negligible	< 0.01 × Normal

dose, not by the number of exposed individuals. However, since there is assumed to be some health risk even at small doses, there should be a requirement to take all reasonable steps to restrict both the individual doses to levels below the action level and the number of exposed individuals. The latter requirement effectively covers the existing optimisation criterion. ICRP takes the scale of action levels developed by Clarke a step further, noting that it is better to avoid a rigid demarcation. It proposes using bands of concern as guidance, which are closely tied to a comparison with natural background radiation dose levels. These are shown in Table 6.2.

No doubt there will be considerable discussion before any revisions are made to a radiological protection system that many in that rather conservative community believe to require little change (e.g. Webb, 2001).

6.4 Relevance of Dose Constraints at the Exemption/Clearance Level

The 0.01 mSv/a figure is the dose level above which exposures may become of concern from the viewpoint of control and optimisation, and, consequently bring a practice or source into the regulatory sphere. In IAEA guidance, this figure is used together with a collective dose figure of 1 man.Sv/a, which, if exceeded, would indicate a need for more detailed optimisation and assessment. Sources and practices giving rise to doses below these figures (so called "trivial" exposures) are of no regulatory concern[14], and are **exempt** from controls. Materials which have been in the regulatory sphere, but have now been treated or conditioned in some way, or whose activity has decayed, such that they could not be used or subjected to conditions which could expose anyone to a dose greater than 0.01 mSv/a can be **cleared** from any further regulatory control.

[14]Care must be taken in using the concept of negligible or trivial levels of dose. If not properly presented, it can appear that regulators are ignoring or trivialising what are regarded as possible hazards, in the public mind. As mentioned earlier, this has caused problems for the NRC in the USA.

Cleared materials would typically be recycled or disposed of along with other industrial or domestic wastes in a landfill. Some materials may be cleared unconditionally (they can be used for any purpose) while others may only be cleared conditionally. This is usually because only a limited number of potential "fates" has been considered as feasible: typically if a material is to be land-filled, then doses for release scenarios from a landfill will be the only ones that have been calculated (and the material may not be radiologically suitable for recycling or other direct uses).

The types of waste which might be appropriate for clearance are limited to low-activity wastes from industrial, research or medical laboratories, some hospital wastes and some decommissioning wastes. Some of these may be generated within the nuclear industry. HLW is obviously not a candidate for clearance. However, an obvious question concerns the situation where a HLW repository, at a particular site and with a specified EBS, is calculated never to give rise to doses above the 0.01 mSv/a level. This is analogous to the circumstance where type scenarios for conditionally cleared low-activity medical wastes disposed of in a landfill never to give rise to doses above the 0.01 mSv/a level.

Could HLW be conditionally cleared from further regulatory control if the condition was that a particular site, EBS and closure system were used? The answer would seem to be no, as there are always likely to be some scenarios (probably intrusion scenarios) which would give rise to higher doses. In addition, the clearance concept is developed for materials in the biosphere, and not for doses resulting from materials isolated far from the biosphere.

Nevertheless, final closure of a repository and termination of institutional control over the site would *de facto* mean that the wastes have been removed from further regulatory control. The disposal system will have been designed to optimise radiation doses to be ALARA as well as meeting dose constraints. However, as discussed above, constrained optimisation of a repository would only aim at showing that individual doses from natural exposure pathways were in the 0.3 mSv/a region. As ICRP 46 points out (paragraph 87), there is no analogy to exemption for potential exposures, such as those that could result in the future from a deep repository. However, it indicates that scenarios which give rise to doses below the exemption level might justifiably be disregarded in the decision-making process.

This issue is largely a matter of confusion in terminology and usage. It should also be recognised that the IAEA has given only limited guidance on applying clearance to certain classes of materials (IAEA, 1998b). The clearance/exemption levels and guidelines were never intended to apply to HLW repositories: they were intended for removing very low specific activity wastes, which intrinsically present trivial radiation hazards, from the rigours of regulatory control, so that they can be readily managed in the everyday human environment. This is why the 0.01 mSv/a level is set so low. It is obviously inappropriate to apply them to much more concentrated wastes, where scenarios with high doses will always be conceivable. In this context, the controllable dose approach may offer the most sensible answer, with constrained optimisation aiming at ensuring that doses from the repository do not exceed a fraction of a millisievert.

A caveat on this matter concerns the current deliberations of the OSPAR Commission on releases of radioactivity to the sea (see Appendix 1). Although this agreement is currently a purely North Atlantic area matter, it may be a precedent for international approaches to controlling marine pollution. If the OSPAR approach described below were adopted widely, it would be of direct concern for a repository releasing directly (possibly even indirectly) to the marine environment.

The Sintra agreement in 1998 laid out the OSPAR Strategy with regard to Radioactive Substances; the agreement has far-reaching effects for the nuclear industry, as well as potential implications for radioactive waste repositories. The stated objective of this strategy is:

> "...to prevent pollution of the maritime area from ionising radiation through progressive and substantial reductions of discharges, emissions and losses of radioactive substances, with the ultimate aim of concentrations in the environment near background values for naturally occurring radioactive substances and close to zero for artificial radioactive substances".

Such emissions would include those from a waste repository. This wording will be developed and interpreted before any real application, and this will be taking place over the next few years. The definition of the term "close to zero" in the OSPAR strategy is of particular concern. One alternative being investigated is a "confidence margins" approach: confidence in the level of precision with which environmental concentrations can realistically be measured in terms of their incremental addition to natural radiation, and their behaviour modelled. In this approach, it might be required that, by 2020 (the OSPAR target date), the contribution of ongoing routine radioactive discharges to concentrations in the marine environment should be indistinguishable in practice from the natural variability in environmental concentrations due to past discharges.

These developments may eventually prove far-reaching. The dose levels being examined are low; at the clearance/exemption level of dose. OSPAR is also considering how to assess impacts on non-human biota, which may also become a general trend in radiological protection (see Section 6.5.1) and may need to be reflected in future setting of standards. In particular, the approach to natural variability and confidence limits might usefully be considered within a performance indicator in future HLW standards. Both these latter issues are developed further in the next section.

6.5 Other Performance Measures

The limitations of dose and risk with increasing time into the future have led to other repository performance measures being suggested. An underlying feature of several of these measures in comparison with natural systems. As discussed earlier in this chapter, comparisons with radiation doses from the natural background (see Box 7) underpin the concepts of radiological protection and the latest ICRP considerations make connection more apparent than it has been previously. However, more direct

Box 7: Natural Radioactivity

We live in a naturally radioactive environment. This is of fundamental importance for the establishment of principles and standards, as the "natural background" to which we are each exposed has been used for many decades as the obvious benchmark against which to judge safe and acceptable levels of "incremental" radiation exposure. The principle is that exposures that add only fractionally to those that we incur everyday, and which themselves show a wide variation from place to place, should have tolerable health impacts.

Natural radiation sources are divided into cosmic radiation, external terrestrial radioactivity and internal radioactivity from within our own bodies. Cosmic radiation comprises largely protons and alpha particles that originate in outer space, mostly from within our own galaxy or near the surface of the sun. The high-energy particles interact with atoms and molecules in the air, producing other nucleonic particles, the ratios of which thus vary with altitude above Earth's surface. Cosmic radiation is responsible for a world average effective dose of 0.38 mSv/a. Some radionuclides are also produced by these interactions in the atmosphere, notably ^{22}Na, ^{14}C and ^{3}H, which can then become incorporated into surface waters, soils and living material. These cosmogenic radionuclides contribute only 0.01 mSv/a to our average worldwide effective dose.

Natural radionuclides in soils and building materials are responsible for external exposures to our bodies (from the gamma-emitting members of the ^{238}U and ^{232}Th decay chains and ^{40}K) that average 0.48 mSv/a. Individual natural radionuclides such as ^{40}K, and the parents of the decay chains, originated at the formation of the universe. They were incorporated into primordial terrestrial materials, where they have resided ever since, slowly decaying and being cycled through other materials. They are only present today because their half-lives are typically of the order of billions of years. Internal radiation exposures come from the same group of natural radionuclides, that we inhale in dust, ingest in food and water, or which reside within our bodies. For example, ^{40}K is an isotopic component of potassium, forming bone, and (at 0.17 mSv/a) is responsible for over half of our internal radiation dose (setting aside radon inhalation, discussed later).

By far the largest component of natural background exposure comes from inhaling ^{222}Rn gas and its short-lived decay products (^{218}Po, ^{214}Pb, ^{214}Bi and ^{214}Po). Radon is a short-lived member (half-life 3.8 days) of the ^{238}U decay chain and is produced in rocks and soils, and building materials made from them. ^{220}Rn (sometimes called "thoron", as it is a member of the ^{232}Th decay chain), is a lesser contributor to overall radon exposures, together with its short-lived daughters. Rocks with elevated concentrations of uranium and radium, such as granites, and soils with large concentrations of radioactive minerals such as monazite, produce the most radon. Radon can be a concern in unventilated indoor spaces constructed of some types of material, or on certain rocks and soils. Radon and thoron together contribute 1.25 mSv/a to our average radiation exposure.

Overall, worldwide natural exposures to radiation would be expected to result in an effective dose to the majority of people in the range 1–10 mSv/a, with a central value of 2.4 mSv/a. Sizeable population groups are exposed to annual doses of 20–30 mSv in some areas. Individuals or small groups might be exposed to doses several times higher than this, if they live at spots with exceptionally high natural backgrounds (e.g. where mineral sands or radioactive ore bodies occur near the surface, particularly if there is some industrial working and concentration of these materials). For example, the population of the city of Ramsar in northern Iran, is exposed to doses as high as 260 mSv/a (Ghiassi-nejad et al., 2002). The individual source proportions of the worldwide average exposure to natural radiation of 2.4 mSv/a are shown in the diagram below. The source of most of the information in this section is UNSCEAR, 2000.

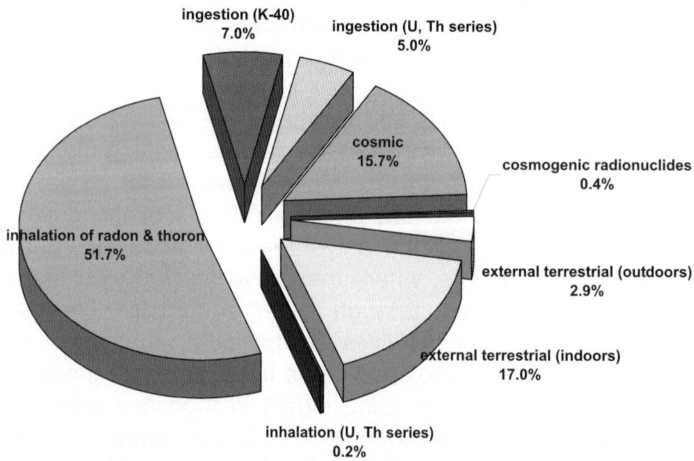

comparisons with processes in natural systems can also be made, as a complement to dose/risk based measures.

Natural systems that exhibit some of the important characteristics of waste repositories, or that evolve in response to the same physical and chemical processes that govern repository behaviour, are often called "natural analogues" (see Box 6 and Miller et al., 2001). An obvious example would be a uranium ore body located some hundreds of metres below the surface. Qualitative arguments, based on natural analogues, can be used by waste disposers to build up a robust case to support compliance calculations. This approach has been used in several national programmes, although it is usually applied at a subsidiary level, rather than being presented as the backbone of a safety case.

More specifically, measures that are being suggested to complement dose or risk include:

1. Fluxes of radionuclides from the repository into the environment, and concentrations in relevant sub-systems, averaged over long time periods.
2. Comparisons with fluxes and concentrations of natural radioactivity (and other natural toxic substances) in the environment: essentially a comparison with natural background radioactivity or chemotoxicity.
3. Toxicity indices for the waste itself.
4. Sub-system criteria, such as container lifetimes and radionuclide fluxes through specific engineered barriers.

There is also interest in setting up measures, perhaps related to those above, which will ensure protection of the overall natural environment (in addition to people).

At the outset, it is important to note that the list above does not identify **alternative** measures. These should be regarded as **complementary** indicators which will add to the information base used by decision makers. The challenge is to identify how to calculate them, how to use them and how to adjust the weight given to all the measures at different times in the future.

Work within the IAEA (BIOMASS, 1999a; Wingefors et al., 1999) suggests categorising the *non-dose/risk* indicators into primary and secondary groups:

- **Primary indicators** are closely related to estimates of radiological impact on people or the environment and thus are direct indicators of the safety of the repository. They can be compared with criteria or reference information which are independent of data derived from a safety assessment, and would include measures 1 to 3, above. For short-lived radionuclides only, IAEA suggests that some sub-system criteria 4 would also be considered a primary indicator (e.g. container lifetime). Protection of the environment as well as people would need to be indicated by this primary group.
- **Secondary indicators** can only be compared with sub-system criteria or reference values which are themselves derived from safety assessments, and all fall within group 4, above. For example, safety assessments can quantify the importance of longevity of engineered barriers, release rates from the waste form and fluxes through engineered barriers. These can all be secondary indicators of performance, but alone are inadequate indicators of safety. Clearly, they are not independent of the results of standard safety assessments.

Quantitative secondary indicators were widely used some years ago in the USA, where, for example, the USNRC stipulated permissible fractional release rates of radionuclides from waste packages and across arbitrary boundaries in the far-field (USNRC, 1983a). These were introduced because it was believed that, in the highly litigious framework of the US programme, it would be easier to demonstrate compliance with a set of fixed quantitative criteria related to engineering and the geosphere. It soon proved to be extremely hard to do this, as similar uncertainties to those that surrounded calculations of dose or risk as a measure also applied to estimating these other, sub-system parameters. It was also widely recognised that fulfilling all the sub-system requirements did not guarantee overall system safety and, conversely, that overall safety could be achieved without fulfilling them all. Quantitative secondary indicators thus appear to have no place in standards and regulations. This is recognised by the USNRC in its more recent regulations (USNRC, 2001; see also Appendix 2).

However, they do constitute useful working measures or design targets that can be used by repository developers in carrying out and assisting in the understanding of the basis of a safety assessment, or in system design. An example of this latter application is setting a design criterion for a container lifetime of 1000 years; an example of the former application by implementors is in presenting the fraction of radionuclides retained in specific components of the disposal system (e.g. canister, near-field, rock) at various times after disposal. This has been used, for example, by Nagra (Switzerland), JNC (Japan) and AECL (Canada; see Fig. 6.3).

Such presentations can be a highly transparent means of communicating a safety concept to the public. It is thus suggested that secondary indicators should form one component of the suite of performance indicators that should be presented by the

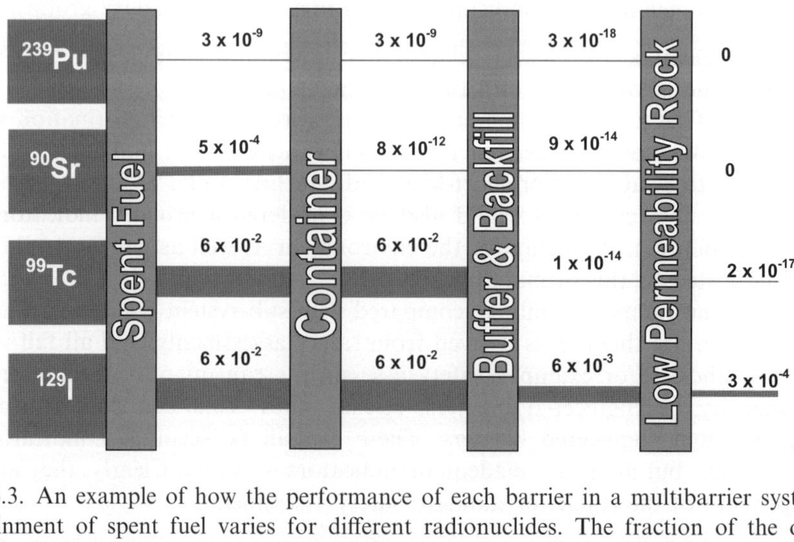

Fig. 6.3. An example of how the performance of each barrier in a multibarrier system for containment of spent fuel varies for different radionuclides. The fraction of the original inventory of some selected radionuclides that escapes past each barrier in turn is indicated. ^{239}Pu is largely immobile within the fuel, ^{90}Sr decays almost entirely within the container, a significant amount of ^{99}Tc enters the buffer but is stopped there, while a small fraction of ^{129}I escapes into the biosphere. This type of presentation is a useful way of illustrating barrier functions (after AECL, 1994).

implementor, but that no quantitative requirements are placed on them in standards and regulations (USNRC, 2001).

IAEA work also identifies a third group of **performance indicators**, which describe necessary or acceptable intrinsic properties of components of the repository; requirements so that they function properly within the design and safety concept. They might also be termed *technical specifications* (IAEA says *criteria*), and include items such as waste loading per package, metallurgical properties of containers and density and composition of buffer material. Again, we do not consider these to be appropriate matters for regulatory standards.

Given the previously mentioned limitations of secondary indicators in a regulatory, rather than an implementor's context, the remaining discussion is thus restricted to what the IAEA is now calling primary indicators. Before discussing these, however, the issue of non-human radiological protection (to which primary indicators might also be applied) is considered.

6.5.1 Radiological Protection of the Natural Environment

The framework for radiological protection of people works in the opposite way to that used for protecting humans from most other environmental hazards. Commonly, society tries to adopt measures for protecting the environment, since that is then assumed to protect humans also. In radiological terms, it is always stated that, if people are adequately protected, then the environment is adequately protected. ICRP 60 states (paragraph 16) that:

> "... the standard of environmental control needed to protect man to the degree thought desirable will ensure that other species are not put at risk. Occasionally, individual members of non-human species might be harmed, but not to the extent of endangering whole species or creating imbalance between species".

It has been argued that this is a largely empirical belief, unsupported by any substantial body of evidence. It may not always hold true, particularly when all time and space scales are considered: for example, possibly high doses to benthic fauna from historic seabed disposal, but only minute doses to people (Pentreath, 1998). There is, for example, no equivalent to the Sv for other species, although there are several studies of the effects of different dose rates on various fish, mammal and bird species. Consequently, there is a growing movement towards including some form of protection of non-human species within the regulatory framework. The IAEA has produced a discussion report (IAEA, 1999b) that notes that there is now sufficient information to be able to move forward to serious consideration of an approach to protection.

Pentreath (1998) has suggested two possible approaches, based on work carried out by the IAEA:

- Continue to base protection on estimated doses to people, but make supporting calculations of other environmental impacts to demonstrate that the ICRP 60 belief is appropriate for other species;

- Determine a number of benchmark dose rate limits for representative, reference species (flora and fauna) and demonstrate, using appropriate models, that these would not be exceeded by the release rates calculated for the repository (e.g. a benchmark value of 10 mGy/day has been suggested for aquatic organisms).

Although there is a large amount of information on radiation effects on the environment (e.g. UNSCEAR, 1996 and 2000: see Box 7, on natural radioactivity), there is no internationally recognised way to use this to reconcile and develop the two approaches, and certainly no agreement on how to incorporate such considerations into regulations for waste disposal.

Based on the UNCED Convention on Biological Diversity (part of the UNEP Rio Conference of 1992; UNCED, 1992), Larsson & Sundell-Bergman (1999) suggest that a general regulatory objective could be formulated in terms of:

- Protection of biological diversity: identifying real or generic critical populations in different ecosystems and ensuring that these populations are not significantly threatened by releases of radionuclides
- Protecting biological resources by ensuring that critical organisms with economic or cultural value are not threatened.

This type of "diversity-resources" approach is reflected in Swedish regulations (SSI, 1997), which also note that biological effects of releases should take account of isolated populations, endemic species and threatened, or particularly valuable species. Smith et al. (1999a) propose that future regulations might also wish to make a closer comparison with non-nuclear waste disposal and environmental protection approaches, including the use of "no observable effect" and toxicity or ecosystem susceptibility based approaches. Some of these considerations may be more appropriate for routine discharges within the broader realm of nuclear activities, rather than for HLW disposal. Releases from a deep HLW repository into the biosphere are estimated to be at such low rates that, unless some significant biological re-concentration mechanism exists, dose rates in the wider environment are likely to be extremely small.

That there is still a long way to go before a coherent, integrated approach to protecting both people and the environment is agreed is illustrated by the very general findings of the most recent study group (NEA, 2003) and by the ICRP opinion that dose limits should not be set for non-human species (MacLachlan, 2003). Within future standard-setting it is thus considered to be premature to make any quantitative or specific provision for protecting non-human species. However, it may be appropriate to include a general requirement for broader environmental protection, for example by requesting the implementors to demonstrate that dose rates in the environment are not grossly unevenly distributed between species and that no specific species or population is threatened in terms of its viability.

6.5.2 Fluxes of Radionuclides from the Repository into the Environment

The basis of this indicator is the calculation of fluxes of radionuclides from the repository into the accessible environment or biosphere (aquifers, lakes, rivers, seas) and their concentrations in relevant sub-systems. The reason for using this is that the processes determining such fluxes are more stable over long periods of time than are the additional biosphere quantities that would be needed to translate the radionuclide fluxes into dose estimates. At present, this criterion is included within Finnish regulations for spent fuel disposal, as a performance measure to be used for the longer term: after "several thousand years". The requirement is to show that the quantities of radioactive substances migrating from the repository, averaged over long time periods, shall be less than radionuclide-specific constraints. Such constraint levels should be at values such that, at their maximum, they are not higher than the level of impacts arising from natural radioactive substances and, on the large scale, the radiation impacts remain insignificantly low.

To date, the Finnish implementor s (POSIVA) safety assessment work (Vieno and Nordman, 1999) has selected 10,000 years as the time after which to begin these types of calculation, and 10,000-year periods as the averaging intervals for the estimates. They also use radionuclide-specific constraints, discussed with the Finnish regulatory agency, STUK. These were subsequently revised and eventually formalised in the 2001 Finnish national regulations (STUK, 2001) as follows:

- 0.03 Gbq/a for the long-lived, alpha-emitting radium, thorium, protactinium, plutonium, americium and curium isotopes
- 0.1 GBq/a for ^{79}SE, ^{129}I and ^{237}NP
- 0.3 GBq/a for ^{14}C, ^{36}Cl and ^{135}Cs and for the long-lived uranium isotopes
- 1 GBq/a for ^{94}Nb and ^{126}Sn
- 3 GBq/a for ^{99}Tc
- 10 GBq/a for ^{93}Zr
- 30 GBq/a for ^{59}Ni
- 100 GBq/a for ^{107}Pd and ^{151}Sm

The STUK regulations note that these constraints apply to activity releases that arise from the expected evolution scenarios and which may enter the environment only after several thousands of years. They can be averaged over 1000 years at the most. The sum of the ratios between the radionuclide-specific activity releases and the respective constraints shall be less than one.

An obvious question is how to derive such constraints. Releases have to be assumed to occur into a given volume of an agreed environmental "compartment" in order for the release rates to have any meaning in terms of concentrations and consequent health impacts. One approach to this would be to use estimated concentrations of each radionuclide (Bq/l) divided by ICRP specified Annual Limits for Intake (ALI) for that radionuclide (ICRP, 1991b). Alternatively, overall release constraints could be set simply as a multiple of the number of ALIs, that is as ALI/a rather than Bq/a.

In addition, the time intervals for the calculations need to be short enough not to "dilute" peak releases by including periods of low release (e.g. as a result of radioactive decay). Thus, there needs to be some definition of how such calculations are carried out.

6.5.3 Comparisons with Fluxes of Natural Radioactivity (and Other Toxic Substances) Through the Environment

There is a clear overlap of the time-averaged repository flux indicator discussed in the previous section, with the second primary indicator dealt with in this section: comparisons with concentrations and fluxes of natural radioactivity (and other toxic substances) in the environment. However, many of the radionuclides of interest in waste (and identified in the Finnish work cited above) do not occur naturally, so release constraints must be based on the calculated radiological impacts of each non-natural radionuclide compared with the impacts of natural radionuclides moving through the same environmental media (groundwater, rivers, etc). Aggregation of these results for all radionuclides would indicate whether certain levels would give impacts higher than those arising from natural radioactive substances.

The most studied natural radioactive fluxes are those of uranium and thorium (and their daughter radionuclides), and of carbon-14 and tritium. There is significant experience on looking at such fluxes, at various scales, from global, to national to repository (Miller et al., 1996). It was noted that fluxes due to surface processes are orders of magnitude greater than in groundwaters and surface waters, essentially because the former are in solid form (eroded particulate material). Solid material (especially in soils and sediments) can give rise to contact exposure, as well as being a source of radon.

One approach would be to restrict comparisons to natural fluxes in groundwaters, the principal pathway for repository releases. In some contexts, when high uplift rates may cause eventual exposure of waste material at the earth's surface, it would also be worth considering comparison with fluxes of eroded natural radionuclides. Again, the time interval over which to make such comparisons would need to be agreed. Nagra carried out such an assessment for the Wellenberg and the Oberbauernstock potential repository sites for the low-level wastes, both in the Alpine region of Switzerland.

The previous discussion raises the issue of the extent to which natural fluxes can also be used to compare the impacts of natural disruptive events (see Chapter 8). Uplift and erosion was mentioned as an obvious possibility, where there is some analytical experience. However, it should be feasible to apply this approach to other scenarios, the most obvious of which would be the impacts of volcanism, which is of concern in some national situations (e.g. Japan and the USA). Volcanic activity, and its peripheral effects, mobilise large amounts of radioactive and toxic materials. For example, huge amounts of fluids can be transported through the crust in and around active volcanic centres, in both gaseous form and as heated groundwaters.

Some work has been done in the USA for the Yucca Mountain Project on the impacts of direct volcanic intrusion into a repository (DOE, 2001a), but scenario analyses tend to concentrate more on peripheral effects some distance from volcanic centres, such as thermal perturbations and changes to groundwater flow and composition. Enhanced natural fluxes (e.g. of radon and other toxic gases) and hydrothermal chemical changes to local groundwaters and potential drinking water supplies would provide useful yardsticks for comparing enhanced releases from a repository for such scenarios. This is an area, which has not, to our knowledge, been explored. Even so, it must be noted that the principal impacts of such events on society will have little to do with a radioactive waste repository (see Chapter 8).

Comparisons with natural radiation impacts could to be carried out at a site-specific level, but it may also be possible to set generic values, typical of a broad, even worldwide range of geological environments and rock types.

A final flux comparison could be with the health impacts of other natural toxic substances moving through the environment (Pb, Cd, Cu, As, Zn, Se, etc, as well as toxic gases, mentioned in the context of volcanism). In some cases, the toxic elements might also be present in non-active repository releases, so there would be a direct comparison. The other approach is simply to regard these as other environmental toxins which have health consequences additional to radiation impacts. Calculation of the health impacts of the amounts of natural lead or cadmium in water might be a useful illustrative material, but they are unlikely to be sensible components of regulatory standards.

6.5.4 Toxicity (or Hazard) Indices

"Radiotoxicity" is a quantitative concept that is usually applied to the waste itself, rather than to released radionuclides. The radiotoxicity index is the specific activity (of a particular radionuclide) divided by the "toxicity", expressed in terms of the radionuclide-specific ALI, which itself takes account of the fact that the radiation emitted by different radionuclides has different biological effects. In this sense, the radiotoxicity of a given quantity of a radioactive nuclide is equivalent to the number of individuals who could receive their maximum allowable annual intake of that nuclide, were the material to be so distributed. The overall toxicity of the waste at any time can be obtained by aggregating these indices in a Hazard Index. Alternatively to ALIs, maximum permissible body burdens (MPBBs) or maximum permissible concentrations in drinking water (MPC_W) might be used as the comparator (see, for example, Liljenzin & Rydberg, 1996).

6.6 Return to Nature — an Approach to Standards for the Long Term

Throughout this chapter, we have observed that radiological protection is based upon the yardstick of natural environmental radioactivity. Section 6.5 showed how various approaches are being used to explore natural fluxes and concentrations as complementary indicators of safety. In this section we describe an approach to setting waste containment objectives and associated standards that take better account of the fact that, ultimately, we effectively return the waste in a repository to nature, after we can no longer rely on the containment functions.

As discussed in Section 2.2 and by Chapman (2002), a properly implemented repository at a good site can contain long-lived wastes such as spent fuel and HLW until they have decayed to levels of hazard commensurate with natural uranium ore deposits: a few hundred thousand years. The strategy of "concentrate and contain", which provides exceptional protection for extremely long periods, leaves, as the inevitable by-product, something akin to an ore body. In the tightest, highest containment rock formation, the longest-lived natural series radionuclides will remain exactly where they were placed, for geologically long periods of time. It could be argued that a rich uranium ore deposit was not originally present in this or that location country, by the time that the ore deposit can have any kind of impact in the environment, perhaps a million years in the future, we are interested only in an environmentally benign global solution.

Beyond this natural "cross-over" time there is a strong case, based on the parallel with nature, on society's real expectations and on sensible use of resources, for saying that we have done enough. There is no logical or ethical reason for trying to provide more protection than the population already has from Earth's natural radiation environment, in which it lives and evolves. It is a scientifically tenuous position to argue that additional protection (e.g. down to a few microsieverts of exposure) can be provided so far into the future and that this can be ensured by regulations.

What would this approach mean in terms of the protection provided by a deep repository? A system of simple "*time-graded containment objectives*" could be envisaged for the designers and siters of deep repositories, which would provide the following broad levels of protection:

Level 1: Zero impact: total containment of all activity in the repository for about 1000 years, the period when it is at its most hazardous. This is already much longer than the period that is of greatest concern to society.

Level 2: For the next one (or a few) hundred thousand years, any releases through natural mechanisms to give rise to doses that are below the range of natural background radiation.

Level 3: After this time, the hazard being equivalent to natural radiation hazards, there is no further containment objective: doses may be envisaged in the range of those from natural background radiation.

In more specific terms, Level 2 would be the period to which the spirit of current radiological protection principles, described earlier in this chapter, could be applied. It is a period over which meaningful estimates of future system behaviour can be made. The performance measure appropriate to this period, and to the approach advocated here, would be to have reasonable expectation that any impacts (assuming the same biosphere as today) are less than the worldwide variation in normal background radiation (excluding the highly variable radon contribution). If these variations are taken to be around 10% of the average dose from natural radiation (2.5 mSv/a) then one arrives at a figure close to the 0.3 mSv/a proposed by the ICRP. This would correspond to what ICRP (2001), in its basis for discussion on new recommendations, is currently calling Band 3: Low Concern.

In the period beyond one (or a few) hundred thousand years (Level 3), it must be recognised and accepted that man cannot be expected over indefinite times to do much better than nature. The potential exists for natural uranium ore deposits, or spent fuel or HLW repositories, to give rise locally to doses that are higher than the global average for natural radiation, particularly if they are eventually eroded in the near-surface environment. However, people exist today in many locations where doses are tens, even up to a hundred times (e.g. Ghiassi-nejad et al., 2002) higher than the average. Thus, a repository is not providing, globally, a novel source of exposure[15] and does not at these long times represent any unusual anomaly in the global environment. It might be expected that the eventual redistribution of residual radioactivity in the environment by erosion and other natural processes should be indistinguishable from regional variations in concentrations of natural terrestrial radioactivity in near-surface rocks, soils and waters: with "regional" taken in the broad sense of, for example, Europe or North America — perhaps even globally.

ICRP 81 notes that judgement is required in optimising protection, and says that the Commission's view is that, provided that the appropriate constraint for natural release processes is met, that reasonable measures have been taken to reduce the probability of intrusion and that sound engineering, technical and managerial principles have been followed, then radiological protection requirements can be considered satisfied. For the Level 2 period, the objective of the approach outlined here is the same, as is the natural release performance measure. The suggestion for Level 3 interprets ICRP 81 views on the progressively decreasing relevance of numerical performance measures with time much more broadly, but, it is believed, pragmatically.

The intention of the approach suggested here is to recognise the value of an excellent concept (geological disposal), to accept that with this excellence come unavoidable ultra-long-term implications, to do the very best that we can in

[15]It is also worth remembering that this variability is being found to be greater than was envisaged at the time of the deliberations that underpin current radiological protection principles.

engineering the solution and to provide optimum protection where society wants it most, and where we have the most responsibility. In doing this we are fulfilling our ethical responsibilities in the most practical way and going far beyond provisions for the future made in any other field of human endeavour.

6.7 Using Performance Measures in Setting Standards

It is a challenging task to extract from the extensive and complex list of topics addressed in this chapter a focussed and concise set of recommendations on formulating protection standards for geological repositories. We believe that the following key issues need to be considered:

- All types of performance measures, including dose and risk, should be considered as elements in a spectrum of safety indicators. Standards and regulations should allow this spectrum to be used in an integrated, holistic way in reaching decisions, rather than relying heavily on only one or two indicators.
- Individual doses are at the core of radiological protection. However, PA calculations of doses after a period of several hundred years can only be regarded as guiding estimates. They are not precise predictions of health detriment. The latest ICRP advice is that dose constraint values of a fraction of a mSv (e.g. 0.3 mSv) should be considered for regulatory purposes, for a "whole" repository (i.e. possibly a repository complex).
- Individual risks derived from the doses by simply using a health effects conversion factor are an almost equivalent alternative. The arguments for specifying risk rather than dose are that no changes are needed if advances in knowledge lead to a change in this conversion factor and that comparisons with other risk activities are easier.
- In estimating individual doses, reference biospheres should be set up to account for potential future environments that may exist where releases may occur. These should be used to identify relevant exposure groups, within which calculations should be made for an average adult individual at the high end of the range of exposure.
- Constrained optimisation, based on quantitative estimates of prospective radiation doses, can be used to help make siting and design decisions. However, the usefulness of the dose constraint as a hard compliance indicator over extended time frames is doubtful and it should only be applied in conjunction with other, sometimes qualitative, criteria. Concerning optimisation, the emphasis should be on ensuring that reasonable measures are taken to reduce future doses. It should not be confused with a formal optimisation process, less still with cost–benefit analysis.
- The latest ICRP guidance could be interpreted as indicating a shift away from using only radiological risks as an indicator, towards a disaggregated indicator of dose-plus-likelihood. The latter element, likelihood, could be expressed

qualitatively (e.g. for intrusion or disruptive events and processes). The two elements could be considered separately in reaching decisions. Such an indicator could be calculated as if it were being used in the normal ICRP dose constraint framework, but a more qualitative approach would be taken to applying it.

- Compliance with regulations must be based on reasonable assurance rather than absolute demonstration, and will require latitude of judgement.
- We consider collective dose estimates with respect to disposal of long-lived wastes inappropriate.
- Quantitative values of dose at the clearance/exemption level (0.01 mSv/a) are inappropriate for use in long-lived radioactive waste disposal regulations.
- If the trend of international thinking continues to develop along current lines, new standards and regulations should consider including a general but non-quantitative requirement for protection of the overall environment (more than just people).
- Any new set of regulations should consider the primary indicators concept being developed within IAEA. These should include natural fluxes compared to repository fluxes, as well as overall time-averaged repository fluxes and environmental concentrations of radionuclides. The issue of when to apply these and how to weight them is intimately linked to the matter of timescales, discussed in Chapter 5. Other indicators, such as toxicity and sub-system secondary indicators are primarily useful tools for the implementor and presenter of information, but they may also provide further useful information for the regulator. It is not considered appropriate to set quantitative criteria for any of these (primary or secondary), but they should all be presented as a suite of indicators.

To conclude, we have suggested (Section 6.6) a system of "time-graded containment objectives" which put the focus on the fact that a repository is effectively being returned to nature after closure and its radiological significance is most appropriately compared to natural systems after some hundred of thousands of years.

Chapter 7

Siting Requirements within Standards

This looks like it could be a suitable site...

This chapter discusses the types of guidance or requirements on repository siting that it might be appropriate for regulators to give to an implementor. The programme and procedures of the implementor in the siting area can be subjected to external requirements at four different levels. These are:

1. *International requirements.* International legal agreements that affect waste disposal. Principally, these are regional (e.g. EU) legislation on environmental

impact assessment (EIA), and the various agreements and protocols on disposal of waste in the sea, which could affect coastal repository locations (see Appendix 1).
2. *Policy and programmatic requirements.* These can lay down the process to be followed. This could include, for example, defining the level of involvement of Government, planning authorities, regulators and the public; the number of candidate sites to be considered; the timetable to be adhered to.
3. *Specific site-selection requirements.* These might cover, for example, looking at a range of host-rock types and geological environments before choosing a preferred host, or avoidance of resources to ensure that site investigations address specified issues.
4. *Site characterisation requirements.* These might specify which data to collect for safety assessment, the QA measures to be applied and measures to avoid negative impacts on site characteristics.

With respect to international requirements (1), there is a trend towards requiring full environmental impact assessments of siting options, evaluating all environmental effects, not only those concerned with repository safety. The specific legal agreements related to disposal of wastes at sea could be taken as affecting the siting of repositories in coastal areas. These are discussed in Appendix 1. National policy and legislation are normally used to set policy and programmatic requirements (2) and sometimes specific site-selection requirements (3). Regulatory bodies (both nuclear and non-nuclear/planning authorities) may also specify requirements (3), and they are normally responsible for any rules or guidance on site characterisation (4).

The involvement of different political and regulatory bodies at each stage in the siting process varies widely from country to country and in some countries there is a confusing and unhelpful mixture or requirements in these different areas. This chapter is principally concerned with the regulatory and planning aspects and only considers programmatic and legal requirements in national laws in terms of how they might affect the framework within which regulations lie.

We begin by looking at international technical guidelines on siting from the IAEA, go on to look at the types of siting regulations that are found in a range of national regulatory standards and then briefly discuss the most important or contentious topics in siting.

7.1 International Guidelines

7.1.1 IAEA

The IAEA most recently established broad siting guidelines for a deep waste repository in 1994 (IAEA, 1994a). This "Safety Series" report was intended both for implementors and for regulatory agencies involved in developing standards, criteria and specifications. Some key points of guidance with respect to site selection are abstracted below:

1. A suitable site can be identified either by narrowing down from a group of candidates or by objective evaluation of one or more designated sites.
2. Existing nuclear sites (or adjoining land) may be worthy of special consideration owing to the potential benefits in relation to reduced waste transport requirements.
3. It may be possible to solicit volunteer sites from communities or owners.
4. It is not essential or possible to locate the best possible site.
5. The approach to assessing safety should be similar, regardless of how the site is chosen.
6. At each stage of a siting process, societal, ecological and legislative issues should be evaluated and addressed according to national policies.
7. The regulatory body should be kept informed of and involved in decisions at relevant stages of the process.
8. A QA programme should be established early in the siting process.
9. Throughout the siting process, data should be collected, presented and archived in a standardised fashion: this needs to be established early in the siting process.

The IAEA identifies four stages to the siting process that have been widely referred to in national programmes:

- **Conceptual and planning stage:** in which an overall plan is developed for the site-selection process, and potential geological environments are identified using available data. Key decision points need to be identified. Screening guidelines should be developed that would enable a repository to be located so as to match national performance criteria and socio-economic, political and environmental considerations. The funding, resources, safety assessment and regulatory background should be established.
- **Area survey stage:** which identifies areas that may contain suitable sites using the screening guidelines developed in the previous stage. This may be a stepwise screening of a region of interest, or gathering regional data for previously designated sites of interest. This stage would also tend to use available data. National laws and regulations need to be considered (e.g. with respect to sensitive environments, national parks, water resources, etc.). These are generally already existent and well-defined and, in many programmes, no specific regulatory decisions would be needed at this stage.
- **Site characterisation stage:** involves the investigation of one or more sites to demonstrate that they would be suitable from the safety and other viewpoints. One or more sites would be carried forward to the next stage after a detailed safety assessment. This would need to be thoroughly reviewed by the regulatory agencies who would decide whether the site(s) is/are likely to be suitable and whether the final stage of site confirmation would be likely to result in a license

application. Again, the regulatory objectives would need to be defined in advance for this step.

- **Site confirmation stage:** in which detailed studies lead up to confirmation of the site and a licence application. This stage would also involve preparation of an **environmental impact assessment (EIA)**, depending on national laws. The regulatory agencies would have to evaluate all the data and analyses in order to approve a licence to construct a repository. Further down the road, separate licence applications are likely to be required to place waste in the repository and to close it, following the evaluation of more data from the construction and operation phases.

The IAEA also provides a set of general **site-selection guidelines** that can be used as one component (along with safety, feasibility, social, economic and environmental considerations) to develop practical national guidelines, should these be considered necessary. It is noted by the IAEA that this is not a complete set, nor in order of importance and they should not be applied in isolation. Their use should also take national limitations into account. Their are reproduced in precise form below. The report also identifies typical data needs to show that each guideline has been met. These are not reproduced below.

- The geological setting should be amenable to characterisation, should have geometrical, geomechanical, geochemical and hydrogeological characteristics that inhibit radionuclide transport and allow safe repository construction, operation and closure.
- The host rock and repository containment system should not be adversely affected by future dynamic processes of climate change, neotectonics, seismicity, volcanism, diapirism, etc.
- The hydrogeological environment should tend to restrict groundwater flow and support waste isolation.
- The physicochemical and geochemical characteristics should limit radionuclide releases to the environment.
- Potential future human activities should be considered in siting and the likelihood that such activities could adversely affect the isolation capability should be minimised.
- Surface and underground characteristics should allow optimised infrastructure design in accordance with mining rules.
- The site should be located such that waste transport to it does not give rise to unacceptable radiation or environmental impacts.
- Site choice should mean that the local environmental quality will not be adversely affected, or such effects should be mitigated to an acceptable degree.
- Land use and ownership in the area of the site should be considered in connection with possible future development and regional planning.
- The overall societal impact of developing a repository at the chosen site should be acceptable, with beneficial effects being enhanced and negative effects minimised.

7.1.2 European Community

In addition to the IAEA guidelines, further international guidance is provided by a European Community study published in 1992 (CEC, 1992). This report notes the parallel requirements of nuclear safety, radiological protection and planning, environmental, socio-political regulations. It emphasiscs that the paramount issue is that the *total disposal system* should meet the requirements set by radiological protection standards and that the site's function is to contribute to the overall safety of the system. In other words, the safety aspects of siting should not be regulated upon in isolation from total system performance.

The report suggests siting criteria for a deep repository that cover:

- geological stability;
- hydrogeology;
- chemical and geochemical properties;
- mechanical and thermal properties;
- depth and dimensions of the host rock;
- presence of natural resources.

Generally, these identify much the same points as the IAEA guidelines discussed above and are not highly prescriptive. However, the stability criteria are unusually quantitatively prescriptive. They propose that:

- "... tectonic movement should not be expected to occur (or to induce significant phenomena) before, e.g. 10,000 years, evaluated at regional levels and forecasted from present trends and evidences of events in the past. More generally, the site should be deemed to be stable as long as necessary according to the safety assessment."
- "Seismicity shall be low. Its acceptable level depends on the option and the site, but it shall be shown that tectonic movements are not expected to reach Level 7 of the Richter scale (or an intensity of IX–X in the modified Mercalli scale)."

These proposals are, of course, made in a European context and the first appears to have been based on criteria suggested in France, in 1987 (Goguel, 1987). It is interesting to note that none of the recent national regulations in Europe appears to have used these two quantitative CEC criteria, although the present French regulations (see below) have picked up the 10,000-year stability figure.

7.2 National Regulatory Guidelines or Standards on Siting

The IAEA guidelines have been adopted to varying extents in national regulations. Table 7.1 gives examples (in brief, rather than in full text) of the main components of national regulations concerning siting. Much of this information is contained in NEA (1997). It can be seen that the level of detail adopted in each country is highly variable.

Table 7.1

Country	**What the regulations say about siting**
UK (Environment Agency)	• Regulator expects to be involved at an early stage of the repository development process • No regulations concerning site selection, but the regulators would expect to be consulted on whether the site is suitable for purpose • Several requirements concerning site investigation (implementor should develop a reasoned approach, account for potential disturbance of site by investigations, show the environment is adequately characterised, understood and capable of analysis to support a safety case, that site characteristics can act to limit radionuclide migration, that future environmental change is accounted for and that potentially valuable resources in the area are identified and the extent to which the site might be disturbed as a consequence is considered)
Spain (Nuclear Safety Council)	• A set of "general siting criteria" has been established by the NSC. In outline, these stipulate that: • Shape and dimensions of host rock body should be adequate • Host rock should have lithology and depth consistent with types and amounts of waste • Site should be characterisable • Stable tectonic conditions are required and active structures and potential faults must be avoided • Site should have no adverse impacts from seismic ground motion • Areas with abnormally high geothermal gradients or evidence of recent volcanic activity should be avoided • Site features should be entirely favourable for waste isolation • Site characteristics should act to limit radionuclide migration • Geotechnical properties should not unfavourably affect disposal • Repository should not be adversely affected by surface processes • Preferable location in areas of low population density, accounting for future growth • Avoid areas of known foreseeable future natural resources: need for repository should be balanced against future need for resources • Repository should not adversely affect local environment
Finland (STUK)	• Conditions should be stable or predictable for at least several thousand years, with subsequent geological changes being estimable • Minimum requirements for siting would be to avoid: proximity of exploitable natural resources, exceptionally high rock stresses, seismic or tectonic anomalies and exceptional conditions (e.g. lack of reducing buffering capacity) for groundwater chemistry

	• Favourable structural and flow properties allowing adequate distance between containers and potential fast flowpaths • Depth of several hundred metres to mitigate impacts of surface events and render intrusion difficult
France (Nuclear Installations Safety Directorate, Ministry of Industry and Trade)	• Fundamental Safety Rule No III.2.f (Ministry of Industry and Trade, 1991) defines a set of largely qualitative siting criteria, which in brief are as follows: ESSENTIAL CRITERIA • Possible modifications of initial site conditions due to glaciation, earthquakes or neotectonic movements shall remain acceptable for safety. Stability, with limited, predictable evolution, shall be proved for a period of at least 10,000 years • Must have very low permeability and a low hydraulic gradient in the host formation and a low gradient should be sought for surrounding formations IMPORTANT CRITERIA • Site should have favourable mechanical and thermal properties • Geochemical properties should be described quantitatively to enable radionuclide transport analysis • Depth should not allow significant effects of erosion, earthquakes or following current human intrusion. Depths should be below 150–200 m • Known or possible "exceptional" interest sites for natural resources should be avoided DESIGN CONSTRAINTS • A large enough block of rock without large water conducting fractures with vaults away from medium-sized fractures (crystalline rocks) or in a formation with no large heterogeneities and distant from aquifers (sediments) • Thermal and access excavation effects should not significantly affect containment
Sweden	• There are no siting stipulations from the regulatory agencies and hence no siting regulations • The government supplements basic legislation on the financing of waste management by specifying timetables for siting, the way in which regulators and implementors should interact during siting and the numbers of sites that should be carried forward at each stage. It has also appointed a special co-ordinator to facilitate siting discussions
Canada (AECB)	• Host rock and geological system should have properties such that releases are retarded • Little likelihood that site will be exploited for natural resources • Geological system should be capable of withstanding stresses without significant structural deformation or fracturing • Dimensions of host rock should allow repository to be located deep underground and away from major discontinuities

It can be seen that only the Spanish regulatory guidelines are a close match to the details provided by the IAEA report. Other countries have chosen to be less prescriptive, preferring instead to stipulate only broad "common sense" factors that should be accounted for in siting, and basing their regulations on the end point of actual performance: ensuring that a proper safety case is made that meets radiological performance measures.

The most recently started efforts on developing siting criteria are those of the German AkEnd group (AkEnd, 2000, 2001). In 1999, following widespread controversy about the two potential sites chosen for deep geological disposal in the 1970s and 1980s (Gorleben and Konrad), the German government set up a group to develop new criteria. This group attempted, in a multi-year process, simultaneously to:

- derive criteria for identifying favourable geological settings;
- propose a process for applying these criteria with maximum participation of all stakeholders, including the public;
- keep the option open for having volunteer communities participate in the process.

7.3 Key Contentious Issues in Siting

This section draws attention to specific aspects of siting which have led to intensive discussion in the waste disposal programmes of various countries.

Progressive technical selection process or volunteering Originally, international recommendations and national programmes tended to aim for a purely technical procedure progressively narrowing in on a final preferred site, with each step being objective and traceable. Scientifically this was found to be difficult owing to the unavoidable requirement for much subjective judgement when comparing radically different types of criteria. It was also sociologically naive in that it underestimated the strong political and public influences. Volunteering by potential host communities was increasingly recognised as also being a responsible method for siting — with the provision that the sites considered must fulfil the same strict safety criteria no matter how they enter the selection process. In practice, it is perfectly possible to run a screening process and a volunteering process in parallel if this basic condition is adhered to.

Responsibility for the selection of a preferred site for license application In most countries, the responsibility for site selection lies fully in the hands of the implementor. The authorities are involved only at the level of specifying the process and then in judging the acceptability of the site put forward by the implementor. In some cases, however, political bodies involve themselves directly, e.g. the selection of Yucca Mountain as the sole site to be characterised in the USA was done by

Congress. In many cases, even if there is no legal requirement to involve official bodies in the process, it is judged prudent by the implementor to do so on a voluntary basis.

"Best" site or "sufficiently safe site" It is now widely recognised that identifying a "safest" site is not a feasible goal in the siting process. The data and methods do not have sufficient resolution and, in any case, it is impracticable to assess fully all potential sites in a country. A "best" site within a chosen set can be selected using multi-attribute analyses with criteria going beyond radiological safety — but the weighting of criteria will always contain a subjective element and hence be open to debate. The technical consensus view today is that a site should be sufficiently safe in that it can satisfy all safety criteria in a convincingly demonstrable manner.

Stepwise siting procedure and regulatory process Repository implementation programmes should proceed generally in a stepwise or phased manner, including the siting process. The involvement of the nuclear regulatory body varies from country to country. One extreme can be seen in Switzerland, where a permit from the nuclear safety authorities is needed even for exploratory drilling. The other extreme is exemplified in the UK where the Environment Agency had no official role to play even when the implementor applied for permission to construct a deep rock characterisation facility at the potential site for a repository.

Pre-definition of explicit exclusion criteria Exclusion criteria are commonly set at the initial screening phase for potential sites. These can be technical (e.g. presence of natural resources or distance from known geological features) or societal (e.g. avoidance of national parks). Such exclusion criteria are not problematic: even an obviously technically good site, if it has features that are clearly agreed to be unacceptable, must be ruled out. Difficulties can arise, however, at the site characterisation phase when detailed parameters are being measured. There is sometimes a publicly expressed wish for threshold values of characteristics such as rock permeability, degree of fracturing etc. The thought is that exceeding these critical values should lead to abandonment of the site. A danger is that the implementor might prefer to weaken requirements rather than put at risk the investment of time in site selection and characterisation. The scientific problem with any requirement of this type is that the performance and safety of a site never depends upon a single parameter. The performance of the total repository system determines safety. This fact should be made clear as early as possible in order to avoid over-interpretation of isolated parameter measurements during site characterisation and such requirements should never find their way into regulatory standards.

Public participation The degree to which the public participates in selection of a potential repository site is perhaps the aspect of siting that varies most from country to country. Given the increasing requirements worldwide for environmental impact assessments and public participatory processes, it is expected that both implementor and regulator will include interactions with the public in their siting work.

Taking into account the material presented in the preceding sections, appropriate approaches for standards and regulations are discussed below.

7.4 Discussion

As a matter of principle, it would seem appropriate for new standards and regulations to include some general preamble that encourages the implementor to adopt a logical and orderly approach to siting that takes into account "big picture" matters such as:

- ensuring that natural features assuring long-term safety are likely to be found at the site;
- assurance that the site can be characterised well enough to enable a safety case to be made with confidence;
- consideration of broad environmental impacts of repository development;
- ensuring that no obviously better conceptual siting options (from simple, first principles) have been overlooked;
- taking account of local and regional social, planning and development issues[16];
- specifying the participatory process for all stakeholders (i.e. interested and affected parties).

This preamble should make simple reference to the IAEA guidelines, perhaps by using words such as "expecting that the implementor would take them fully into account". There seems to be little point in regulations then going any further and stating what are rather vague, "obvious" and practically useless qualitative requirements, such as those that are found in the Spanish regulations (e.g. "the host rock should have lithology and depth consistent with the types and amounts of waste to be disposed").

We would suggest that specific requirements should be stipulated *only in areas that would assist the process of finding and proving a suitable site* and making it transparent to all parties.

It will also be important to decide whether regulations on siting are to be split between agencies. Some of the items bulleted at the start of this discussion section

[16]The two final bullet points in this list might be more appropriately dealt with in national policy and law rather than in regulations. It is important, however, that they are not overlooked, hence their inclusion here.

are clearly in the field of general environmental impacts rather than being in the radiological protection or nuclear regulatory field. Given an opportunity to develop new standards, it would clearly be advantageous to have an integrated set of regulations that covers all environmental impacts of siting, thus avoiding the "double jeopardy" problem of the implementor having to satisfy two independent sets of rules (or three, if a government also stipulates detailed siting process, by law). An integrated set of regulations would place most emphasis on production of a satisfactory EIA, in which radiological safety assessment would only be one component.

Bearing in mind the above points, topics which might usefully be considered for an integrated set of regulations include:

- Defining a set of clear **exclusion** guidelines that minimise the risk of certain (e.g. tectonic) scenarios so that they do not need to be accounted for, or they can be accounted for in an agreed manner. These exclusion guidelines would be aimed at removing obviously unsuitable regions (e.g. close to volcanic centres or along active fault traces) from the siting process and also simplifying the licensing discussions between implementor and regulator by eliminating or reducing the importance of specific scenarios. Similarly, a requirement to site in geological structural domains which are well known to have little or no resource prospectivity would assist with the human intrusion issue. These two matters are discussed in more detail in other parts of this book, with respect both to exclusion guidelines in siting (Chapter 8) and to appropriate regulatory performance standards (Chapter 6).
- Stipulation that the implementor should get guidance on acceptable siting strategy from government policy-making agencies. The type of requirement that might emerge from these agencies is that there must be a choice of sites, up to a point where at least two have been characterised in parallel, using equivalent approaches, to a level of detail that allows realistic comparison via a formal EIA. This provides both a sensible flexibility in the national programme, and reassurance to the public that the process is not able to be manipulated easily to produce a result at a preferred site, no matter how it performs. It would be important for the regulations to define the required content and structure of the EIA. For example, in Europe, this basis is established by the European Union EIA Directive (European Commission, 1997).
- Definition of a set of milestones for siting work. These would identify what is expected of the implementor at each step of the site-selection process, but not necessarily the duration of each step or how the targets are achieved. The milestones could define the documentation that should be submitted at each of the siting stages that may be stipulated. For example, if a national survey of suitable regions and geological environments is considered an appropriate first step, then the regulations could require a comparative safety assessment only at this stage, with no EIA. If the next step is to be limited investigation of a small group of sites, then outline EIAs might be stipulated for each site, including

preliminary site-specific safety assessments. If the final step is to be a comparison of two or more sites in detail, then a full EIA would be required.

An effective set of standards and regulations would make it clear that siting decisions will be made on the basis of an EIA which includes a dominant element of long-term safety assessment. Performance measures for the assessment would be defined elsewhere and would simply underpin the siting assessments. Because safety would be judged on the basis of total system performance, and because some scenarios would be explicitly treated in the siting regulations (volcanicity, intrusion: see above), it is not appropriate to include further "sub-system" requirements. Thus, quantitative or semi-quantitative stipulations about distances from aquifers, repository depth, type of host rock, geological stability, etc., would be unnecessary, because their importance would be swept-up and accounted for in the total system performance analysis.

Chapter 8

Natural Disruptive Events and Processes

Safety assessment studies of deep repositories are often based on a central "reference case" description of the "normal (undisturbed) evolution" of the disposal system (see Section 2.3). Although terminology varies from one national programme to another, the approach usually taken is to evaluate the "as-designed" performance of the repository within such a central case, and to consider separately processes and events that could disturb this progressive evolution, frequently as part of a "scenario analysis". The disruptive events are often further sub-divided into natural and human-induced categories. This chapter examines the first, natural, category and the next chapter looks at the potential impacts of people on a repository.

For most European waste disposal programmes, natural disruptive events and processes are largely those driven by climate change. Long-term geological stability in Europe means that, as far as neotectonics are concerned, assessments generally only need to evaluate seismic impacts on repositories, and, perhaps, progressive land uplift and erosion. More tectonically active countries (e.g. USA and Japan) need to look at a wider spectrum of potentially disruptive events. In this chapter we consider how disruptive events might be evaluated and how regulations on repository siting and safety might address the issue. The suggestions we make at the end of the chapter are inevitably speculative. No organisation has yet applied a systematic approach to disruptive events in a repository site-selection programme (i.e. to compare sites or siting areas), although NUMO, the Japanese waste agency, is now establishing one.

Natural disruptive events are commonly perceived as being obviously dangerous to the safety of a deep repository. Furthermore, in some parts of the world there is a high probability that some type of disruptive event will occur in the region of any repository site chosen, at some time during the next 100,000 years or so that is typically considered in safety assessments. A systematic approach to characterising natural disruptive events and evaluating their impacts thus has the objective of providing a credible and easily understood analysis that should, for properly sited and designed repositories, show that such events:

- will not add immediate, acute radiological hazards to other hazards that might already be caused by the event;
- will not damage the repository so much that it will cease to contain the wastes properly and constitute an unacceptable hazard long after the event has occurred.

Making this case means meeting the quantitative and qualitative requirements of regulations, satisfying public concerns that are unlikely to be allayed by reference to statistics, taking all flavours of expert scientific opinion into account and reassuring political and other decision makers that all stakeholders have been consulted and credible arguments presented.

Each of the target groups will need to receive information in different forms, which must be internally consistent and rooted in sound scientific arguments. We begin by looking at a possible approach to building the underlying scientific case and applying it credibly in the siting process for a deep repository. We then look at how the implementor might present the results of their assessment to the regulator (or, conversely, what the regulator might sensibly ask for). The final part of the chapter considers how this technical basis can be interpreted for use by the public and decision makers.

8.1 Identifying Disruptive Events

The appropriate approach to identifying what may constitute disruptive events is by means of analysis of features, events and processes (FEPs). This is normally carried out as part of the initial systematic description of the repository and its environment. The objective of such analyses is to determine all of the FEPs that describe the system and could affect its evolution and performance. The FEPs are normally identified by a process of iterative "brainstorming" by broad groups of experts covering a range of technical fields, although there are now several comprehensive FEP lists that can be used as a starting point for an environment or region-specific study (e.g. NEA, 2000c).

Screening of these FEPs allows many to be set aside as being of little or no importance or relevance because their impact is negligible, their probability is extremely low (e.g. less than one in one thousand million per year) or their other consequences dwarf any impacts on a repository (e.g. large meteor impact, which also has an extremely low probability).

Following these steps, it is common to deal with most of the remaining FEPs by considering their impact on key parameters affecting repository performance when analysing the *expected* evolution of the repository system. Re-running the simulation models for the expected evolution, using the key parameter values as amended by the disruptive events, then allows the overall consequences to be judged. This leaves a residual group of FEPs that defines external events and processes which do not lie within the normal, design basis evolution of the system. In Sweden, the regulatory authority SKI refers to these as external FEPs (EFEPs). Some users call them "scenario-generating FEPs". They will certainly include, and probably be entirely comprised of, all conceivable disruptive processes and events.

It is reasonable to ask whether this rather fundamental activity needs to be repeated in every national programme. It has already been carried out in many national programmes and international inter-comparisons and recent studies have rarely produced anything new or different in terms of scenarios for analysis. A possible approach has two components:

- Carry out a high-level generic study for any new national disposal programme as a once-and-for-all exercise to identify relevant disruptive event FEPs and to see whether there are indeed any disruptive events not considered in the standard lists.
- At a later, concept or site-specific stage, carry out a focussed FEP analysis aimed at specific disruptive processes and events, to ensure that their potential impacts in specified regions/conditions and siting environments are comprehensively identified.

The "standard" disruptive events that will emerge from the first component can be predicted with some certainty now. Neglecting human activities, the main groups of potentially disruptive processes and events are:

- **Climate change effects** (including changes in sea level, glaciation, development of permafrost conditions, variations in precipitation and infiltration, in surface temperatures and in biosphere properties);
- **Uplift and subsidence** (and associated faulting, erosion and changes in drainage patterns, groundwater flow and chemistry, and rock stresses);
- **Seismicity** (including new fault development, fault reactivation, earth movement and groundwater pumping and chemical changes);
- **Volcanicity** (possibly excluding direct intrusion by a major volcanic centre, but including intrusion by minor dyke and sill intrusions, perturbations in rock and groundwater temperatures and geothermal gradient, induced hydrothermal activity, and changes in surface topography by deposition).
- **Flooding, landslips and tsunamis** (all principally with respect to the repository operational phase and the last only in coastal areas).

Last category is not considered further here as it has little impact on long-term, post-closure performance of deep repositories. The exclusion of direct intrusion by a volcanic centre would need to be justified on the basis that the event has considerably greater non-repository based consequences than any likely radiological impacts. This is discussed further below.

The above groups can be categorised into those that constitute progressive *processes* and those that are true *events* of limited time duration. Clearly, most geological events are actually the culmination or the punctuation of some long-term process. An earthquake represents the focussed release of energy that has been stored up over a long time and, possibly, over a large spatial scale. Nevertheless, it is possible to define a group of processes that do not always lead to rapid, catastrophic events but which still need to be assessed in their own right. On this basis, both climate change and uplift/subsidence can be regarded as gradual processes (although potentially giving rise to abrupt environmental changes, which would need to be classed as events).

Recent USDOE definitions (DOE, 2000) are considered useful in this respect:

Events

- occurrences that have a specific starting time and, usually, a duration shorter than the time being simulated in a model;
- uncertain occurrences that take place within a short time relative to the time frame of the model.

Processes

- phenomena and activities that have gradual, continuous interactions with the system being modelled.

8.1.1 Treatment of Potentially Disruptive Processes

Where processes are clearly going to be continuously active during approximately the next 100,000 to one million years, they should be addressed as part of the central "normal evolution" of the disposal system, even though there may be uncertainty about their magnitude, direction and rate. They should form part of the main, quantitative analysis of the reference case, treated in the same way as, for example, groundwater flow or container corrosion, and should *not* be relegated to "what if" scenario analyses. Consequently, disruptive processes are not discussed further here. Instead, we concentrate on disruptive events that are linked to tectonics: principally, volcanicity and sesimicity.

8.2 Taking Account of Disruptive Events in the Site-Selection Process

The IAEA (1994a) recommends that the implementor needs to take account of disruptive events at an early stage in the siting process. If this is done properly, then it may be possible to eliminate some of them from further consideration. They thus need to be integrated into the initial criteria and guidelines used for accepting or rejecting sites and regions during the siting process. For disruptive tectonic events, this could be done in a number of steps, as outlined below:

1. Set-up a baseline tectonic framework for the country concerned This could comprise a broad expert review of the tectonic framework of the region, intended to provide a baseline consensus on key issues such as the current tectonic regime (structural elements, stress and thermal patterns), its stability and likely future evolution. It would be expected to present in a harmonised fashion the most recent models of heat flow, seismic activity distribution, fault activity, crustal displacement (lateral and vertical) and volcanic activity. A key issue to evaluate would be the concept of a "period of stability". In France, regulations require that "geological stability" must be demonstrated over a period of 10,000 years and potentially disruptive processes should be predictable over 100,000 years, which implies some measure of continuity in driving mechanisms. In the USA, the period of geological stability (for the specific site of Yucca Mountain) is considered to be approximately 1 Ma, based on advice from the National Academy of Science (EPA, 2001c). These could both, perhaps, also be interpreted as periods of *"stability of predictability"*. The USA accepts that the *recent* geological record is the most reliable source of data on disruptive event frequency and magnitude and suggests that the Quaternary (the last 2 Ma) is the appropriate time frame for gathering information. This ties in to some extent with the 100,000 year and 1 Ma "stability of predictability" periods suggested above.

2. Eliminate exclusion areas for certain categories of disruptive event: The tectonic baseline is used to exclude areas where disruptive processes and events will contribute considerable uncertainty to the nature and evolution of geological conditions over the next 1 Ma. These areas should be openly rejected from further consideration. The guidelines that could be used for exclusion are areas:

- with markedly elevated heat flow;
- with high lateral crustal displacements;
- with high potential for exposure of wastes due to the aggregate effect of uplift plus erosion);
- with complex stress regimes.

Both Switzerland and Japan provide good examples of how understanding of the regional tectonic framework can be used to establish possible exclusion areas. In Switzerland, quantitative models of potential exposure rates (uplift and erosion) were used to exclude large parts of the country (the Alpine areas) from consideration for siting a HLW repository in hard, crystalline rocks.

In Japan, there is good evidence for assuming that there is an extremely low likelihood of new volcanic activity to the generally south-eastern side of the so-called "Quaternary volcanic front" within the time period relevant to geological disposal. The line of the "front" essentially marks the edge of the region in which all Quaternary volcanic activity has occurred. Its location is defined by the geometry and mechanism of plate subduction, melting and the rise of magma into the overlying lithosphere: it is not an arbitrary line, but one whose nature and location are well understood. First principles suggest that, if the tectonic framework of Japan remains stable for the next million years, a site well to the "inactive" of the front would not experience significant volcanic impacts[17]. In fact, in its top-level exclusion factors, NUMO first uses the more specific option of eliminating anywhere within a 30 km diameter circle around a volcanic centre (NUMO, 2002b). Non-excluded areas, on either side of the "volcanic front", will than be evaluated on their merits.

The alternative to the exclusion approach would be to take a purely risk-based approach. In the case of volcanicity, the implementor would try to show that it was possible credibly to assign probabilities to new activity occurring within any chosen area (events/km^2/a). This approach was used on a localised, site-specific scale in the USA, at Yucca Mountain. Here, the studies have been used belatedly (as the site had already been selected) in an area that turns out to be more volcanically active than first considered. Rather than excluding volcanism scenarios, the USDOE has had to try to demonstrate that the probabilities are low (within the 10,000-year time frame specified in US regulations) and that the consequences are not catastrophic (DOE, 2001a): see Fig. 8.1.

At the end of this stage, some categories of disruptive events would thus be effectively removed from further consideration.

3. Stress and seismic analysis The baseline tectonic information would be developed into a more detailed picture of the stress and deformation regimes in the non-excluded areas, including *major* active features (defined here as showing significant activity during the last 2 Ma). This would be used to make a semi-generic (applicable to large, broad regions) first estimate of seismic event magnitudes and frequencies, including an estimate of the maximum possible event. The evaluation would be expected to look within the "period of stability" (predictability) identified at the outset. Note that this analysis would be revisited in much more detail at the multiple sites stage (and the single site stage), so that it could be regarded here as very much a preliminary, scoping exercise.

4. Concept and formation-specific response models Information on potential siting regions and specific rock structures and formations within them would be combined with the output of stress and seismic analysis. This would be used to estimate both the rock mass and the hydraulic responses of each region/formation to recurrent seismic events, single large events and steady deformation (uplift). For the seismic model, an approach based on evaluating the impacts on near-field rock properties of "maximum possible" earthquake magnitudes and their possible distances from a repository would be appropriate (e.g. LaPointe et al., 1997, 1999). At the region-specific level, this should be regarded as a first, largely generic application, with a moderate expectation of level of detail.

[17]Even outside the area of any future activity there may be some small effects (e.g. ash cover) that need to be considered, at some level, in a comprehensive safety assessment.

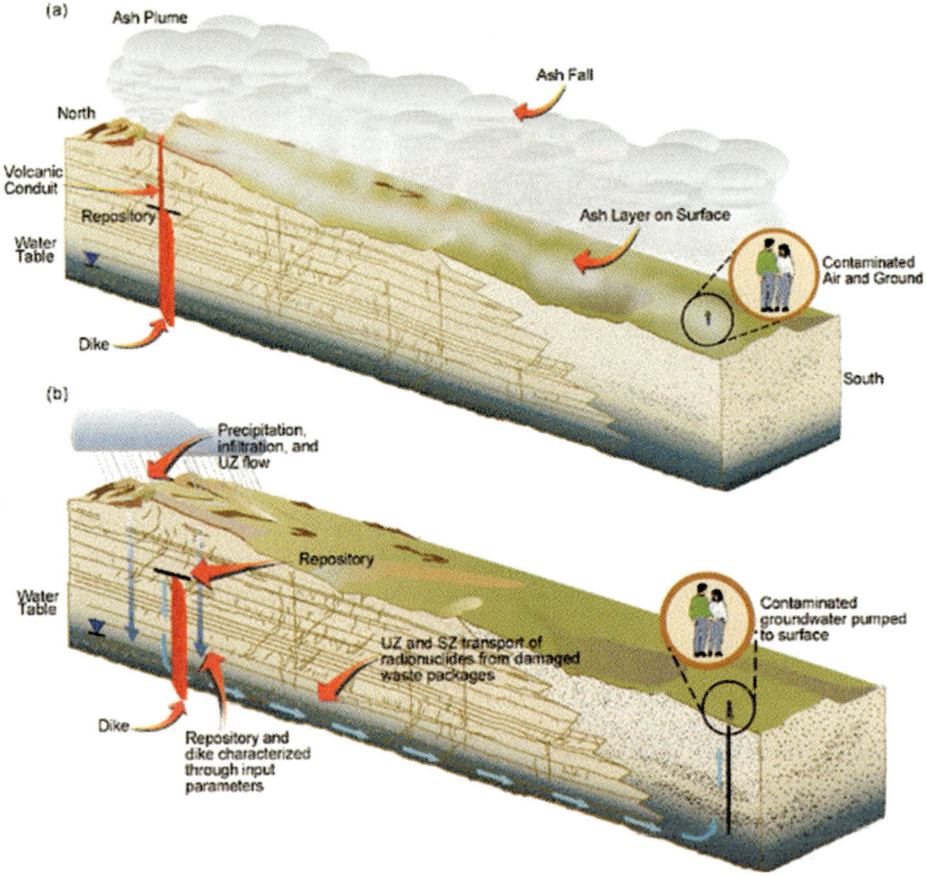

Fig. 8.1. Schematic representation of the two volcanism scenarios analysed for the Yucca Mountain safety assessment: (a) Volcanic eruption, showing development of an eruptive conduit through a part of the repository, entrainment of waste particles in an ash cloud, and deposition of contaminated ash downwind of the potential repository. (b) Igneous intrusion, showing the intersection of one or more emplacement drifts by an ascending dyke followed by flow of magma in drifts, engulfing and damaging waste packages, and subsequent transport of radionuclides by groundwater moving through the solidified intrusion. UZ = unsaturated zone; SZ = saturated zone (DOE, 2001a).

5. Scoping analyses of impacts The objective of this activity is simply to estimate the impact of the maximum event and maximum exposure assumptions, on hypothetical repositories located at a sensible selection of stylised locations with respect to major active features. These results could then be fed back into the siting process to assist with decisions on the selection of one or more sites for detailed study. This analysis may identify particular sensitivities of certain regions or formations that may make them more or less attractive in terms of predictability or performance.

6. Long-term "fate of system" model An additional input to the transition from consideration of broad regions to specific sites is to evaluate the tectonic "end-point" of a repository. For example, it might be expected that the eventual fate of a repository (on a timescale of a very few million years) will be to be eroded out or deeply buried. This type of information might be a useful contribution to the qualitative performance measures that would be taken into consideration when comparing siting options.

7. Site-specific seismic hazard analysis Once specific sites are identified, then the seismic evaluation would become more detailed and it would be appropriate to carry out a parallel exercise to that used most recently at Yucca Mountain (DOE, 2001a), which is itself based on seismic hazard evaluations for nuclear power plants. The results would feed into comprehensive site-specific safety assessments, with uplift and recurrent small seismic events being considered in the reference case model, and major seismic disruptive events being considered as separate scenarios. It is only at this point that the implementor has to consider in depth how to introduce probabilistic assessments and to organise the performance measures that will be presented. Prior to this, event frequencies would have been considered in a semi-quantitative manner.

8.3 Performance Measures for Disruptive Events: a Disaggregated Dose-Likelihood (DDL) Table

A practical problem exists in deciding the appropriate performance measures to use for assessing disruptive events at different times in the future. It is closely related to performance measures for undisturbed evolution.

The IAEA Safety Requirements for geological disposal (IAEA, 2001b: in draft revision at the time of writing) note that:

> ...protection against less likely events and scenarios is best achieved by siting and design measures to reasonably minimize the likelihood that such events will impact on the repository. Estimates of doses may be presented for illustrative cases and, where feasible, probability or likelihood of occurrence may be estimated. Under some national regulations, the probability and consequence will need to be combined in order to show compliance with a risk target. In other countries, the separate estimates of probability and consequence will inform the regulatory judgement.

ICRP (1998a) lump disruptive events together with "extreme conditions" (with probabilities $<0.01/a$) as "potential exposures". They suggest that such exposures should be evaluated using either aggregated likelihood-consequence (risk) criteria or disaggregated likelihood-and-consequence guidelines (as discussed in Chapter 6). ICRP leans more towards the latter approach. They recognise the "conceptual satisfaction" of doing a comprehensive risk analysis but also point to the difficulty of identifying all exposure situations.

The disaggregated approach (for convenience, called "disaggregated dose-likelihood", or DDL in the remainder of this book) would analyse "representative scenarios" for disruptive events (i.e. the "illustrative" scenarios of IAEA, called "stylised" scenarios by some authors). The relevant comment in ICRP 81 (ICRP, 2000a) is:

> The radiological significance of other, less likely, scenarios can be evaluated from separate consideration of the resultant doses and their probability of occurrence. It should be noted that this approach does not require precise quantification of the probability of such scenarios occurring but rather an appreciation of their radiological consequences balanced against the estimated magnitude of their probability.
>
> ...more information may be obtained for decision-making purposes from separate consideration of the probability of occurrence of a particular situation giving rise to a dose, and the resulting dose.

This then raises the question of the time period over which such evaluations should be made. ICRP suggests two approaches:

- quantitative estimates could be made over, say, 10,000 years, moving then to a range of more qualitative indicators, or;
- the use of both quantitative and qualitative indicators for much longer times, but progressively giving more weight to the more qualitative ones.

Canada, France, Germany and the USA have used a 10,000-year "transition" in this fashion. In Canada, France and Germany, it is a hard transition from a dose/risk criterion to qualitative arguments, in the USA it is a hard compliance cut-off (no regulatory standards apply to performance calculated after this period, even though peak doses out to the end of the "period of stability" of 1 Ma must be presented).

Other countries prefer to use the second approach, using both quantitative and qualitative indicators together. Finland and the UK use the approach almost as ICRP suggests, while Switzerland uses a fixed 10^{-6} risk criterion for disruptive events for all times. The UK is the only system that seems to move towards the same position as ICRP: a disaggregated approach for high consequence — low probability disruptive events.

What can be distilled from this discussion? The following approach seems appropriate if a DDL model is used:

- rather than using a time transition or cut-off, the DDL approach could look at likelihoods and consequences in different time periods that are clearly related to one or more "periods of stability" of a region or site, closely supported by geological arguments;
- DDL in this framework would be used in parallel with other indicators, including non-radiological impacts. The examples discussed below all concern the disruptive potential of seismic activity.

The siting process outlined in Section 8.2 should have identified the magnitudes of seismic events at various distances from a hypothetical repository location that could cause impacts on performance. It is expected that, as in Swedish and US analyses, some kind of "threshold" will be identified in terms of rock or water displacement, below which no significant impacts of a single event are considered possible. The quantitative definition of a significant impact should be determined by sensible scoping calculations of potentially critical changes in the repository system. These could be:

- rock displacement in the buffer-canister region;
- displacement of rock and seal materials at critical sealing points;
- displacement of groundwater by seismic pumping, if appropriate for the formation.

The threshold impact will then be defined in terms of a threshold event magnitude and distance from the repository. Thus, the aim now will be to identify frequencies for events above this threshold and then to calculate their consequences in terms of enhanced radionuclide releases, and non-radiological impacts.

A DDL table can be envisaged (e.g. Table 8.1), showing the information that needs to be presented for an "above threshold" event. This can be carried out at both the concept-specific level and later, at the site-specific level, when a much more precise estimate can be made of event magnitudes and potential locations with respect to active features in the ~100 km around the site. Tables can be constructed for different event magnitudes up to the maximum possible event.

In this approach, impacts and likelihoods are considered in different time frames. The time frames could have both a regulatory importance and be designed to make presentation to the public easier. They have been chosen because they relate both to the public's time horizons and to distinct, identifiable, time-dependent states of an undisturbed repository for which it would be important to explore the impact of a disruptive event. The scale and possibly the nature of the radiological impacts might be different in these periods. The time frames suggested are:

- **Licensing to 100 years:** this covers both the operational period, for which a seismic hazard analysis will certainly be required for licensing purposes, and the immediate post-closure period. It is also the timescale most commonly identified as being of concern to the public — basically the next generations — for whose future well-being people express most concern and in which they are thus likely to be most interested (see Chapter 5).
- **100 to 1000 years:** represents the period over which, for example, a HLW container might (realistically, if unperturbed by disruptive events) be expected to remain intact and during which many fission products in all types of long-lived wastes decay away (the proposed period for the Level 1 containment objective of total containment proposed in Section 6.6).
- **1000 to 100,000 years:** represents the period during which radionuclides (apart from the most mobile) will largely be contained within an undisturbed EBS, and

during which overall waste package radioactivity decays to levels near natural background. It may also be identified as the period of "stability of predictability". It coincides with the Level 2 containment objective period proposed in Section 6.6.

- **100,000 years to "end-point":** The final period is that in which only the longest-lived radionuclides in the waste are of any concern: many will still reside in the waste, but the most mobile may be expected to be widely distributed throughout the disposal system. The "end-point" is where the repository is predicted to be eroded away (uplift & erosion) or buried to an inaccessible depth (subsidence & sedimentation). This is equivalent to the Level 3 containment objective period of Section 6.6.

The impacts of recurrent events below the threshold level also need to be considered. Whilst it is clearly appropriate to treat these in the same way as progressive processes (because their frequency within a time frame will be large), as part of a reference case, it may also be useful to present them in a DDL form. Decision makers and the public may find it useful to see the estimated impacts of below-threshold events, simply because, although their radiological consequences are negligible or non-existent, their non-radiological impacts may be large.

Note that the DDL table in Table 8.1 is divided into an upper "engineering period" region (the first time frame of 100 years in which an engineered response to disruptive events would be available) and the very much longer, subsequent "scientific period", when disruptive event impacts have to be considered theoretically on the basis of scientific knowledge. A valid question is how far the implementor can go towards mitigating the impacts of disruptive events by engineering (and spending money) in either of the two broad periods shown in the table. Certainly, in the first period of 100 years, the implementor must design a repository and its surface facilities to withstand significant seismic events whilst the system is at its most hazardous (open and incomplete, with operational nuclear facilities at the vulnerable ground surface). If several decades of "enhanced" retrievability is an objective, then the EBS and the access works must be designed to allow for this even if damaged. Beyond the "engineering" period, into the three post-closure periods, spending more money on making barriers more robust (e.g. increasing the thickness of bentonite buffer to mitigate against shears affecting containers) would only be justifiable on the basis of an ALARA analysis. Are the estimated risks great enough to enhance the EBS, even if this is practical? Of course, this kind of question is best tested at the concept comparison and multi-site inter-comparison stage, and not left until a site has been selected. If the risks are high enough to necessitate consideration of complex engineering fixes, it is likely that the wrong site has been selected.

8.4 Presenting Information to the Public and Decision Makers

The DDL table provides a good basis for presenting the impacts of disruptive events, provided that the information in each box can be shown in a readily

Table 8.1. DDL table of information that could be presented for, in this example, an estimated maximum possible seismic event (e.g. M_W 8.5–9) "within range" of a site. The values shown are entirely hypothetical, for illustration

Time frame (years)	Likelihood of the event occurring in time frame*	Possible physical consequences of the event	Possible radiological consequences of the event	Likely non-radiological consequences of the event
"Engineering" Period				
Licensing to 100	10^{-5}	Major disruption of operations, especially surface facilities Minor damage to EBS (e.g. 10 containers)	Damage to waste storage areas at surface Loss of power or access to repository affects containment Deterministic exposures(10s) Stochastic deaths ~100s	Catastrophic impacts on infrastructure and social fabric within 100 km of site. ~10^4 to 10^5 deaths, including 10s at the facility.
"Scientific" Period				
100–1000	10^{-4}	Minor damage to EBS (e.g. 10 containers) Minor changes in flow paths in host rock	From releases of mobile short-lived FPs into groundwater Doses in 10 mSv range	Catastrophic impacts on infrastructure and social fabric within 100 km of site. ~10^4 to 10^5 deaths

1000 to 100,000	0.01	Minor damage to EBS (e.g. 10 containers) Minor changes in radionuclide release rates	From releases of long-lived mobile iodine, etc. Doses in 0.01 mSv range	Ditto
100,000 to "end-point"?	10	Possibly major damage to EBS if uplifted, or insignificant, if buried	Only from natural series radionuclides still contained in the degraded repository: equivalent to an ore body Doses from well below natural background, down to effectively zero	Ditto

*This may not vary linearly with time, as mechanisms may vary in intensity in different time frames.

understandable fashion. It is suitable both for a regulator's needs and for other "stakeholders".

However, the regulator needs to decide how to weight all these results in coming to a decision. The disaggregated dose-likelihood approach will sometimes reveal calculated doses that are much higher than the usual dose constraint, or even a "relaxed" constraint such as suggested by the ICRP for human intrusion (see Chapter 9). In these circumstances, weighting will need to rely on a value judgement on the likelihood (rather than the dose) component. This type of judgement ought to consider the non-radiological impacts of events (e.g. social consequences, other health risks) in order to give the repository impact some sensible context. It should also consider how society reaches decisions about the risks of other developments that take place in areas prone to stochastic natural hazards. Part of this judgement demands thinking realistically about the timescales that are being considered.

Numerical frequencies such as shown in Table 8.1 are not easily grasped by the public or non-technical decision makers. Likelihood needs to be couched in a semi-quantitative way, or related to something less abstract than future time periods. This is where comparisons with everyday experience, chances and opportunities to which people are exposed may be useful. During each of the time periods shown in the matrix, people have had to cope with immense local and global natural catastrophes and social upheavals, some of which (in the 100,000-year time frame) are believed to have threatened the viability of the human species (see Section 2.4). These points are made simply to illustrate that there will be numerous and greater risks as a result of natural events that people will be subjected to both globally, as Earth's population, and locally, near a repository. These considerations may be relevant when deciding how much weight to attach to the analyses within each group on the table.

There is scope for presenting simple, descriptive information for each of the columns in the table. The only useful comparator for radiation dose is the natural radiation environment. Even though people may not understand the science, the concept of a generally benign environment containing acceptable intrinsic hazards can be communicated. Clearly, there is much scope for comparing hypothetical, stochastic radiological "deaths" to the non-radiological impacts of the final column.

The DDL approach will probably be needed, in any case, to satisfy any likely type of regulatory requirements. However, it will not be enough alone to provide all the information that the public and decision makers need. Decision makers must also take other factors into account:

- No conceivable disruptive event can result in a sudden release into the biosphere of any large portion of the waste inventory, unless such an event is so catastrophic that radiological effects pale into insignificance relative to the accompanying effects (e.g. a giant meteor impact).
- It is not possible to avoid the impacts of some disruptive events completely, for the ~100 ka lifetime of a repository. Many other potentially hazardous facilities are built every year, with the same constraint. Even though they have design lives

of only 25–100 years, they may be intrinsically more susceptible to damage from relatively small disruptive events if they are located at the surface. The contribution of robust design must be explained.

- Would the waste be safer against disruptive events if it were managed in some other way? The impact of seismic events on surface or shallow underground stores must be discussed. Threshold levels of events that could have a significant impact on such facilities are likely to be much lower, and frequencies correspondingly greater.
- Given the massive non-repository impacts of some types of disruptive events on society and the environment, are national resources best spent on trying to reduce already very small impacts of a "disturbed" repository (e.g. by excessive siting or design requirements)? If a case is already robust, then more resources should not be spent on trying to make it even better (the ALARA concept).
- The probability of below-threshold disruptive events could make engineering provisions for "enhanced" retrievability more difficult and may argue for earlier closure of a repository to place it in a passively safe condition as soon as possible.

Chapter 9

Intrusion by Future Generations

The scenario that people might intrude into a repository, with significant radiological impacts for the intruders and possible disturbance of the barrier system, is an inescapable consequence of concentrating the wastes into a small volume of rock. On the other hand, as discussed in Chapter 2, concentration, containment and deep burial of the wastes is acknowledged as being the most appropriate means of isolating them from the biosphere for the long time periods necessary to ensure safety. If we want to take advantage of this solution we thus have no option but to live with the possibility of human intrusion. Nevertheless, there are measures that can be taken to try to reduce the probability of intrusion (and, to some extent, its consequences), at a minimum for times extending out to centuries after repository closure. A key question is if, and how to regulate deep disposal practices with respect to intrusion.

Direct intrusion into a repository lies at one extreme of the range of possible impacts on a repository that could be caused by the future activities of people. This range extends from modifications to the land surface affecting groundwater recharge and discharge, through modifications of nearby underground space (e.g. by mining or tunnelling) to actual disturbance of the repository and the wastes by direct intrusion. This chapter deals only with direct intrusion. This is defined here as intrusion into the repository rock/groundwater volume and the repository structure, resulting in contact with the waste or with materials contaminated by radioactivity from the waste.

Possible future intrusions have usually been categorised as:

- **Intentional,** in full knowledge of the presence of the repository and its probable contents;
- **Inadvertent,** accidental intrusion resulting from a loss of knowledge about the presence of the repository.

For a long time, it has been widely accepted in the waste management community that regulatory requirements should not seek to protect society from intentional intrusion, but should aim to ensure that measures are taken to reduce the probability (and, if possible, the consequences) of inadvertent intrusion, once active institutional control has ceased (normally taken to be somewhere in the period up to 500 years after closure: see Chapters 10 and 11). Intentional intrusion is thus not considered in either regulations or safety assessments.

Inadvertent intrusion into a deep repository could occur as a result of borehole drilling for resource exploitation or scientific purposes, or by underground excavation to extract resources or to utilise underground space. To date, all assessment effort on intrusion scenarios has focussed on the likelihood and consequences of these categories of intrusion. Inadvertent intrusion could lead to direct exposure of people to the waste, or to volumes of rock, water or repository materials contaminated by releases from degraded waste packages. It could also result in the damage of the EBS and repository system such that, even if direct exposure to radioactivity does not occur, the repository system ceases to function as designed, leading to enhanced releases at some time after intrusion takes place. To date, most consideration has been given to the former, direct exposure impacts, and there has been less study of the behaviour of a damaged repository (Smith et al., 1999b).

There is a third possible category of intrusion, which has not been considered before:

- Intentional intrusion out of curiosity, when knowledge of the consequences has been lost. This is called "**naive intrusion**" in this book, and is discussed further below.

The treatment of human intrusion in repository siting and design, safety assessment and regulation, is not purely a technical issue, and due consideration needs to be given to a number of factors, and how to weigh one against another:

- the ethical issue of protecting future generations that may be unaware of the repository or its contents, or less able to cope with an intrusion incident;
- how to assess the importance of intrusion, including the scenarios to be assessed, the types of calculations to be carried out, the consequences and the safety indicators to use;
- if, and how, the results of intrusion analyses should be compared with other scenarios, both disruptive and those reflecting the slow evolution of the repository;
- whether to have separate regulatory standards for application specifically to human intrusion, since the unpredictable element of human intention makes it different in nature to stochastic, natural disruptive processes;
- the timescales involved and the feasibility of estimating meaningful probabilities of intrusion and of devising means to reduce the likelihood of intrusion, or the consequences of intrusion should it occur, and the appropriate level of resources which should be spent doing this.

These factors can be regarded as a group of questions, which the following discussion tries to answer. The conclusions of this chapter present them again, with a summary of recommendations on how they could be handled.

9.1 Types of Intrusion

It has been a common practice to date to evaluate **inadvertent** intrusion using a risk-based approach. Typically, such analyses might focus on an activity that is regarded as both the most likely, and also of relatively high consequence (although for a small critical group): drilling boreholes into the waste or the repository. Estimates have been made of the probability of drilling an exploratory borehole into (for example) a HLW canister, based on current exploration drilling practices and rates in different geological environments, expressed as a frequency/m^2 of repository land area/year. The product of this frequency and the radiological consequences (for the driller and an associated group of geotechnical workers) of contacting retrieved waste in core then provides one measure of risk. Other values can be factored into the probability calculations, such as the likelihood that records will have been lost, the likelihood that the repository would not be detected remotely before drilling commenced, and so on. Some analyses (see below) have also distinguished the risks to drillers/geotechnical workers and those to the public caused by degradation in the performance of the repository following intrusion.

Published results for drilling intrusion for a range of environments tend to indicate low risks (with potentially high consequences for the intruders), but it is clear that the analyses are highly speculative and sensitive to the probability values chosen, particularly the assumptions about future drilling activities and motivations. Entirely logical reasons could be identified for choosing wholly different values, so the results of such analyses can be difficult to defend. The recent passage of the WIPP deep repository for transuranic wastes through the US licensing process was made easier in this respect because the regulator provided specific guidance on the drilling frequency that should be applied to the safety assessment by the implementor (30 boreholes per km^2 over a period of 10,000 years: current drilling rates in the potash-rich siting area were used as the basis for this interesting figure). This simply defined the process for compliance with the regulations, but did not inspire universal confidence in the safety of the repository, as degrees of belief in the likelihood of intrusion at this site by drilling differ considerably.

Analyses of inadvertent intrusion have looked at a range of exposure mechanisms:

- Drilling into waste and direct exposure to recovered material in borehole core: Canada, Germany (Konrad), Netherlands, UK, USA (WIPP and Yucca Mountain), Switzerland (Wellenberg);

- Drilling or mining into a contaminant plume or contaminated aquifer close to the waste or the repository and consumption of water from the borehole, or its use in agriculture: Belgium, Canada, Finland, France, Sweden, Switzerland, USA (WIPP);
- Construction or living on land contaminated with wastes from drillcores: Canada, UK;
- Damage to the repository by adjacent mining or drilling into pressurised fluids below a repository: France, Germany, USA (WIPP);
- The present EPA proposed standards for Yucca Mountain (EPA, 2001c): 40 CFR 197) require calculation of doses that arise via releases down an unsealed, degraded borehole that has penetrated a waste container and gone on to penetrate the underlying aquifer. Calculation of doses to the drillers would not be required.

The *intentional* category of intrusion described earlier includes accessing the repository to enable retrieval of the wastes (see Chapter 4), but it could also include partial excavation of the repository from scientific or archaeological curiosity. Clearly, in the first case, the intruders should be fully aware of the hazards, and of the radiological protection measures required to carry out retrieval safely. However, in the latter case, it is possible that some future generation may have a reduced technical awareness about the nature or hazards of a repository, yet still have the desire and the means to explore what would then be an ancient relic of past human activities. This lies between the two categories: it is neither accidental intrusion, nor intrusion in full knowledge of the likely consequences. As noted above, for the purposes of this discussion, we call this third category "**naive intrusion**"[18].

Even though it is intentional, it is reasonable to try to minimise the probability that naive intrusion will occur, in the same way that we address inadvertent intrusion. One may also consider whether we should try to reduce its consequences, as well as its probability. For intentional intrusion, the only way to protect future generations in the first respect is to ensure that the repository and its hazardous nature is unambiguously marked on the ground and that information about it is disseminated in such a way as to ensure maximum likelihood of the preservation of knowledge (see Chapter 11). It might be argued that, in the case of naive intrusion, the only way to protect against human curiosity is to do the opposite, and leave

[18]Curiosity is an overwhelmingly powerful agent, particularly if it is sustained over generations, and it is interesting to consider that no unusual work of ancient man has ever been left wholly undisturbed once it has been discovered. The corollary to this, which gives some pause for thought, is that, although a repository may be designed to function passively and safely for a million years, it seems unlikely (based on past experience) that it will survive for more than a few hundreds or, perhaps, a few thousands of years without being interfered with, naively or intentionally (with either good, or questionable motives). Although efforts will be made to archive information as securely as possible for future generations, the probability that a deep repository will ever be allowed to demonstrate its long-term containment capacity may be low. This is a good argument for designing a repository — as suggested in the new USA regulations — such that it does not lose significant containment capability when breached by a single hole. This note of realism is never explicitly stated in safety concepts or assessments.

"The usual mumbo-jumbo about a Pharaoh's curse."

the site unmarked. On balance, it is better to hope for continued recognition of a hazard for as long as possible, so a well marked and recorded site appears the most appropriate solution. Chapter 11 contains further discussion on the issue of markers for disposal sites.

9.2 Protection Objectives that Account for the Possibility of Intrusion

If we believe that there is a significant probability of "early" intrusion and we are concerned with trying to reduce the consequences, this might lead to an interesting reassessment of protection objectives, which could, perhaps, be formulated as follows:

1. As the "standard" baseline, design the repository to function passively to achieve doses that are acceptably low at all times, either because the waste stays in place until it has decayed to negligible levels of activity, or because it is ultimately widely dispersed and diluted by natural processes;
2. Place most emphasis on providing maximum protection against natural processes for the first few hundreds or thousands of years after closure, when it can reasonably be assumed (but not guaranteed) that future generations might leave the repository alone;

3. Provide this early protection by a robust engineered barrier system which aims to contain the waste completely for up to 10,000 years, so that even intentional intrusion of any kind is less likely to disperse activity in the environment or lead to the undoing of the underlying safety concept on which the system is based;
4. Locate the repository so as to minimise the probability of inadvertent intrusion should knowledge eventually be lost.

In effect, these objectives place most emphasis on complete containment of the wastes over the first hundreds or thousands of years when they are also at their most hazardous. They thus parallel the Level 1 containment objective advocated (in Section 6.6) on the basis of more general considerations of timescales and feasibility.

In all of the above respects, it has to be accepted that we can only "do our best", using present day technology and a present day model of how people might behave and be motivated in the future. In reality, we can have no scientific basis whatsoever to identify the technologies that will be available to people in a few hundred years time, or of the reasons why they might enter underground space, and it is of little use to invest a large effort in such speculation.

9.3 Possible Regulatory Perspectives

Regulators cannot avoid taking a formal position on the treatment of human intrusion. To date, most national regulations have said very little that is generically definitive or helpful in this respect (although most regulators have stated that sites with significant natural resource potential[19] should be avoided and that deliberate intrusion need not be accounted for). What view should a regulator take of this philosophically difficult issue? Three distinct positions are discussed:

- At one end of the spectrum, there is a case to be made for taking a very pragmatic view that, provided a repository is sited to the best of our ability to avoid features that make the area attractive for future generations to explore, then we have done as much as we can. Intrusion may well happen, but it is a risk that goes with the geological disposal, concentrate-and-contain concept, and it must be accepted. If we have done our best to avoid intrusion then there is no practical value in speculations about the effects of intrusion, or in calculations of their impacts, as the assumptions are exceptionally uncertain and the results cannot be used sensibly in a normal radiological protection framework.

[19] What constitutes a likely prospect for economic resources is a debateable issue. Views will change with time, as exploration and production technology change, as commodity values vary, as materials find new applications and as developing strategic resource demands place new requirements on producers. Some commentators have even suggested that "resource-free" underground space is a valuable resource in its own right. Since it is not useful to attempt quantitative definitions of prospectivity in these circumstances, the intention must simply be to avoid areas containing what, today and for the foreseeable future, would be regarded as obviously potentially significant resources.

- At the other end of the spectrum there is the quantitative regulatory "fix", illustrated by WIPP (see Section 9.1), for a site that is manifestly prone to potential future intrusion as a result of its location in an area rich in potash resources and hydrocarbons. The net effect is the same, in that intrusion is effectively relegated to being a "non-issue", although rigorous application of the first approach above may have made it impossible to select the WIPP site in the first place.
- The middle ground is one where some attempt is made to evaluate the effects of intrusion so that a regulator can be informed about the overall resilience of a site to this factor, as well as to other, natural disruptive events. As discussed below, the present direction of thinking is to keep such analyses to an illustrative level. They may well only be of real utility at the site intercomparison stage, to select a site clearly less prone to intrusion, or one where intrusion impacts would be lower. Once a final site is selected there may then be a case for omitting intrusion from the final licensing analysis.

How, then, might the third, intermediate approach develop? First, there are established principles which should be considered:

- As discussed in Chapter 3, it is well accepted in the basic principles of radioactive waste disposal (IAEA, 1995b) that we should endeavour to provide to future generations the same level of protection against present day actions (i.e. waste disposal, in this context) as we do to current generations.
- The OECD-NEA states that intentional disruptive actions should not be considered in safety assessments, but actions in which the disposal system is inadvertently disrupted should be considered (NEA, 1995b). It is also noted that the analysis of human actions can only be illustrative and never complete.

Intrusion into a repository would not lead to "normal exposures" of people, as defined by ICRP 46 (1985). Existing ICRP advice is that exposures resulting from disruptive events should be treated as "potential exposures" as described in ICRP 64 (ICRP, 1993), and that both their magnitude and probability should be taken into account in reaching waste management decisions. This involves calculating annual individual risk, and a risk objective (for probabilistic scenarios, such as intrusion) of 10^{-5}/a of serious health effects has been in common currency for over 15 years.

However, ICRP recognises that the role of "potential exposures" in risk assessment for long-lived radionuclides is not yet clear. ICRP 81 has recently proposed (ICRP, 2000a) moving away from the separate treatment of "normal evolution" as, over the long term, there will be no single identifiable evolution, only scenarios of possible future states, to which some degree of likelihood can be attached. Repository behaviour can then be represented using risk, or disaggregated, dose-plus-likelihood for each scenario (see Chapter 6 for more detail).

ICRP is currently suggesting (ICRP, 2000a) that, within this revised framework, human intrusion risks should not be considered in the same context as those of other natural scenarios. Specifically, they state:

> "Because the occurrence of human intrusion cannot be totally ruled out, the consequences of one or more typical plausible stylised intrusion scenarios should be considered by the decision maker to evaluate the resilience of the repository to potential intrusion. ... Since no scientific basis exists for predicting the nature or probability of future human actions, it is not appropriate to include the probabilities of such events in a quantitative performance assessment that is to be compared with dose or risk constraints."

This advice is rather different from the conclusions of the 1995 NEA study, which, although not definitive, can be interpreted as coming down in favour of quantitative risk assessment of intrusion using similar techniques to those used to calculate the risks of natural events (NEA, 1995b). NEA does, however, place a strong caveat on thinking of the probability of natural events and the likelihood of human actions in the same light. NEA prefers to term the latter "degree of belief", and suggests using a range of scenarios of such actions that reflect different degrees of belief, rather than a single scenario as a point estimate. Rather confusingly, the NEA then suggests using present day practices to derive, for example, site-specific frequencies of drilling for application to quantitative risk analysis. NEA does note a need for further debate on this issue.

To summarise, NEA appears to come down (albeit rather hesitantly) in favour of quantitative **risk** analysis of intrusion, with a warning about not putting too much belief in the probability values ascribed, whereas ICRP prefers the use of one or more stylised scenarios to calculate acute and prolonged **doses**, with no probabilities (hence risk) attached. The proposed regulations for Yucca Mountain are along the lines of the ICRP thinking.

ICRP 81 (ICRP, 2000a) notes that their previous recommendation of a dose constraint of 0.3 mSv/a for members of the public for the optimisation of protection is not applicable in evaluating the significance of human intrusion. They point out that any protective actions required should be considered during the development of the disposal system. If intrusion could lead to doses sufficiently high to lead to deterministic effects following an acute exposure, or an unacceptable risk of stochastic effects following prolonged exposure, reasonable efforts should be made to reduce the likelihood of intrusion; for example, by increased depth of disposal.

For acute exposures (e.g. to an intruder), ICRP notes that doses less than 0.5 Sv are unlikely to result in serious deterministic effects. For prolonged exposures, if doses are calculated to be less than about 10 mSv/a, intrusion will not require further attention. From 10 to 100 mSv/a, the possibility of reducing likelihood of intrusion will have to be evaluated, considering the magnitude of doses, costs and feasibility. It is interesting to note that the present EPA standards for Yucca Mountain 40 CFR 197 (EPA, 2001c) would require demonstration that non-acute exposure to intrusion-initiated doses are not above 0.15 mSv/a, a considerably lower figure. This figure, however, is not for doses to the intruder but for additional doses to the public caused by the containment of the repository being impaired by a single borehole penetrating the repository.

The issue of possible protective actions that could be taken at the time of disposal, raised by the recent ICRP deliberations, was also evaluated by the NEA. These can be divided in measures that control impacts (once an intrusion into the repository region has occurred) and those which control the initial probabilities of intrusion into the repository. Examples of impact controls are:

- a suitable geometric layout, e.g. ensuring that a single borehole would not penetrate many waste containers;
- backfill materials that are difficult to excavate or drill through;
- waste forms with reduced tendency to form dust if exposed in the open air;
- thick concrete or metal covers above the waste canister deposition zones;
- extra strength waste canisters;
- significantly reduced concentrations of radionuclides in the waste matrix (i.e. a higher glass to waste ratio in HLW).

It is not worth expending resources to protect against low estimated doses that can already be considered to be ALARP. Consequently, regulations should not aim to minimise *impacts* of intrusion by stipulating generically, as a matter of course, that the implementor should include additional anti-intrusion barriers or materials in the repository. If potential doses are already low, such measures may not be justified.

However, if estimated doses are above the 100 mSv/a prolonged exposure dose limits recommended by the ICRP, then adoption of either protection or probability reduction measures should be required in the regulations (and certainly considered in the 10–100 mSv/a range). It is likely that an implementor would opt for the latter, as suggested by ICRP, as it may be much the simplest response to implement. Typical measures designed to minimise the probability of intrusion were identified by the NEA study:

- siting the repository away from areas with currently recognised, subsurface resource potential[20];
- utilising depth of disposal to isolate the waste, as far as practicable;
- conservation and communication of information about location, contents and hazard (see Chapter 11);
- durable physical marker systems on the surface, and/or in the backfilled access tunnels and/or above the waste containers (see Chapter 11).

[20]This measure is a potential problem for a repository located in a salt deposit, where solution mining of what is clearly a potential resource cannot be excluded.

Such measures also have a much greater sense of credibility than speculative engineering "fixes" in the repository, designed to reduce potential impacts. Nevertheless, the regulator would want to be assured that the fundamental design of the repository system was resilient to an intrusion event and that one borehole would not cause the safety system to collapse like a pricked balloon.

If the ICRP recommendations are followed in setting standards and regulations, then there would be a requirement on the implementor to identify one or more stylised intrusion scenarios and calculate their consequences with respect to the dose guidelines discussed above. These scenarios should look at indirect impacts of intrusion on the repository system, as well as at direct exposure of the intruders, and others affected by contaminated materials[21]. It is widely accepted (NEA, 1995b) that any such scenarios should only assume present levels of technology, to avoid unfounded speculation about future society. This is considered acceptable, as the scenarios are only illustrative of how a situation might arise and what the consequences could be. Thus, such scenarios need to be presented with a discussion of possible circumstances and the types and sequences of actions that could lead to exposures. It is important that they also discuss qualitative likelihood within various time frames over, say, the next few thousand years, even if risks are not calculated.

A discussion of context and degrees of belief in likelihood gives some flavour to the scenarios presented, which would help the regulator evaluate the necessary "resilience" of the disposal system, as suggested by ICRP. A qualitative discussion of likelihood is also useful when *comparing* sites, where there may be clear differences in the possibility of certain types of intrusion. Components of some scenarios may lead to doses higher than the ICRP suggested limits, in which case the implementor would need to show that the system had been optimised, preferably with respect to intrusion probability reduction, with these exposure routes in mind.

Clearly, it would be useful to develop an agreed set of stylised intrusion scenarios in the near future, on an international basis. There seems to be no particular reason why these need to be country specific, although they may need to address differences in disposal concept and geological environment.

NEA suggested other activities that it would be useful to carry out internationally to address the intrusion issue. A regulator might require an implementor to get involved in such activities, as they may help to assure a lower probability of early (i.e. within the next few hundreds or thousands of years) intrusion. The concepts, described at length in Chapter 11, can be summarised as:

- development of national and international archives of radioactive waste repositories, conserved at different locations and different societal levels;
- development of a consistent international approach to repository marker systems (as used for many hazardous materials at present).

---

[21]New IAEA suggestions are in preparation. These include discussion of appropriate reference scenarios for human intrusion for all types of waste repository (IAEA, 2001c; 2002).

9.4 Approaches to Setting Standards for Human Intrusion

Intrusion of some form will be impossible to preclude, and may even be likely, within the first few thousands of years after closure of a repository. Speculating on the details of how and when it may occur is generally accepted to be a sterile exercise. It is important for all parties concerned in developing, regulating and endorsing deep repositories to accept this as a risk that is inextricably linked with the disposal solution.

Our prime responsibility is to give future generations the best chance of avoiding blundering accidentally into a hazard of which they are totally or partially unaware and, by so doing, exposing themselves to harm, or so damaging the repository that it no longer functions adequately. Even so, we must accept that there is a significant probability that a small group of intruders might receive high radiation doses. It is not appropriate to treat either these doses, or the risk that these doses will be received, in the same way that we treat other doses and risks, as there is no scientific basis for predicting the nature or probability of future human actions.

Three approaches to a regulatory stratagem were identified, of which two are worth considering in more detail.

1. The first would only be used in the site intercomparison stage, which implies that the regulator would have to give approval of this particular aspect of siting before the final site was chosen. The implementor would be asked to demonstrate that, in selecting a site, they had taken appropriate measures to reduce intrusion probability by avoiding significant resource conflicts when locating the repository, and to give preference to sites where intrusion impacts would be less. There would then be no requirement to consider intrusion in the final licensing of the site, and no requirement for detailed quantitative analyses at any stage.
2. The alternative approach is to carry out some level of illustrative analysis. This might best be utilised, again, at the site intercomparison stage, possibly being omitted at the final licensing stage. The implementor should devise a small group of scenarios for inadvertent or naive intrusion and thereby illustrate the consequences of such intrusions (including impacts on the performance of the repository), using dose estimates. These should be accompanied by a discussion of the circumstances under which such a scenario could arise, the groups of people that could be exposed and the sequences of actions that would lead to exposure. From this information, the implementor might wish to comment on their degree of belief in the likelihood of such intrusion occurring. The implementor would also provide a description of the measures taken to reduce the probability that intrusion could occur. The magnitude of exposures calculated would be reviewed and the circumstances under which people might be exposed. The regulator would then come to a decision based on the balance of:

   - calculated exposures resulting directly or indirectly from intrusion;
   - measures taken by the implementor to reduce intrusion probabilities;

- the performance of the repository under conditions undisturbed by human intrusion, within an evolving natural environment (the "mainstream" output of PA).

In making this judgement, it must be expected that some calculated intrusion exposures could be higher than values suggested by the ICRP. No intrusion-related dose constraints would be set in such regulations.

In either approach, the regulator would stipulate a minimum requirement of actions to reduce intrusion probabilities, including the keeping of records and the marking of the site. They would also, at an early stage in the regulatory process, review the site-selection procedures to ensure that the implementor had taken all reasonable care to avoid areas with higher likelihood of intrusion.

Chapter 10

Monitoring and Controlling a Repository before and after Closure

A geological repository is likely to be operational for many years or decades. A considerable amount of monitoring activity will take place both before and during that period, covering many aspects of the stability and behaviour of the repository. Monitoring of some type is likely to be required by society for a long period *after* closure as well, and it will also be necessary to consider controls over land use into the distant future (see Fig. 10.1). Making sure that the repository will not be interfered with inadvertently, as discussed in the previous chapter, has two components:

- physical control or supervision of the land containing the repository, so that access or land use can be restricted;
- maintenance of some form of institutional memory of the repository and its contents for future generations.

The term "post-closure institutional control" is frequently heard in the context of near-surface repositories for short-lived wastes, where physical control involves limiting access to the site and continuing some level of maintenance of the closed repository. It is usually intended that, after this control period, the site can be released from all forms of control, as radioactivity concentrations in the predominantly short-lived wastes have decayed to levels such that potential health impacts from any exposure scenario would be below concern (i.e. would not give rise to individual doses greater than 0.01 mSv/a: see Section 6.4). Typically, this period is stated to be about 300 years (about 10 half-lives of the typical main sources of activity, ^{90}Sr and ^{137}Cs).

Clearly, for long-lived wastes in a deep repository, such low exposure levels will never be achieved for some scenarios (e.g. intrusion) in any period that is reasonable for control. A deep repository is designed to prevent significant releases to the surface, especially when the wastes are at their most hazardous, and to isolate the

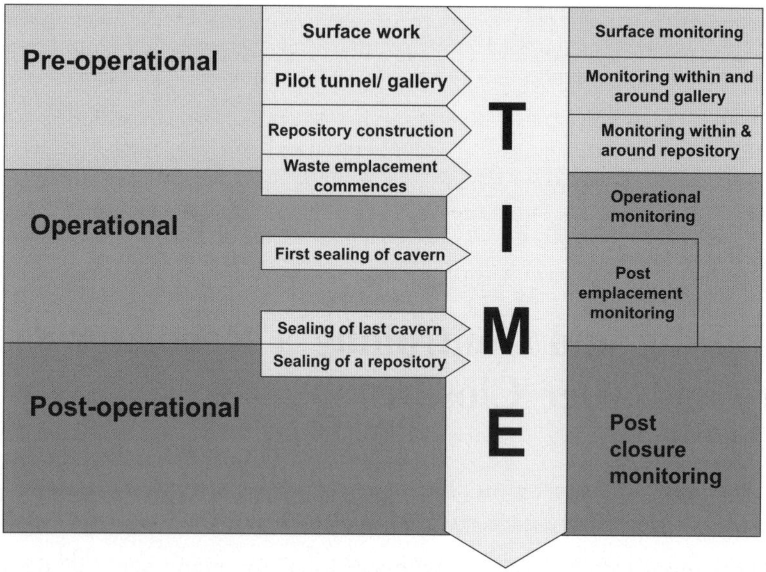

Fig. 10.1. Typical monitoring stages related to activities during the development, operation and post-closure periods of a repository.

wastes fully from the activities of people: a siting programme should have endeavoured to locate the facility in a region where deep drilling for resources will be unlikely. Consequently, most programmes assume that, after closure, sites will not **need** to be subject to long-term physical control.

The fact that active monitoring is not technically necessary does not preclude it being implemented as a confidence-enhancing measure. However, for a properly implemented repository, the more important, safety-relevant institutional requirement is for the maintenance of records of the location and nature of the repository for as long as possible. Thus, as the programme moves from the operational period through and beyond closure, the nature of physical control will change the initial, direct control of the land around and above the repository. It might be expected that this control will progressively be removed, and the emphasis will shift to ensuring proper archiving of information on the repository. Chapter 11 addresses the question of preserving information on the repository in order to reduce the probability of intrusion after the active control phase is ended. The current chapter concentrates on the issue of direct measures for monitoring and controlling the site.

The picture is complicated by a number of factors:

- a national retrievability policy may require that the access points to a sealed repository remain under long-term control until there is no further demand for maintenance of retrieval capability;
- a spent fuel repository may have to be subject to nuclear safeguards controls for an unspecified time after closure and, depending on location and design, this may

require some form of control, or at least surveillance, of the site: it should be noted that ensuring ease of retrievability and maintaining strict safeguards present conflicting requirements;

- there is no agreed international position on the significance of "release from control" of the wastes in a deep repository (effectively amounting to "clearance" in ICRP and IAEA parlance) at a time when the wastes and their potential radiological impacts clearly would not meet quantitative IAEA clearance criteria if they were in the human environment;
- Although technical analyses indicate that monitoring a repository after closure will reveal little, societal demands and consequent national policies on long-term monitoring of the environment of a sealed repository may require long-term observations. For this, some measure of control over the surface in the vicinity to allow certain types of monitoring, and there may be continued liabilities to the implementor, or the state, to maintain any such programme.

The discussion on retrievability in Chapter 4 looked at the first factor, but there is currently little international precedent to draw upon for the second and third. The final factor introduces two important matters: the type, level and purpose of monitoring that might be demanded in national regulations at any time during the life of a repository and how a site might be marked and recorded to maintain knowledge far into the future. Monitoring is the focus of the present chapter, while Chapter 11 examines the preservation of knowledge about a repository.

Much of the text in this chapter is adapted from a recent IAEA commentary on monitoring (IAEA, 2001a), which was developed with considerable input from the authors of this book.

10.1 Monitoring Objectives

It is a widely accepted principle of disposal that, once wastes are isolated in a sealed repository, then the long-term safety of the disposal system should require no further actions on the part of future generations. One component of this principle is that ensuring the long-term safety of the disposal system should not be dependent upon the continual monitoring of its behaviour. This is not to say that people in the future might not want to monitor the repository in some way, but its safety would not depend on them being willing, or able, to do so. The system should be designed to be intrinsically and passively safe. Why, then, is monitoring of interest in a repository development programme?

First, it must be appreciated that there are many steps on the road to the eventual closure of a geological repository. Monitoring could have several useful technical applications in the potentially extensive period up to this point, which may be possibly many decades after the start of a repository development programme. Second, there may be a valid, and an extremely strong, social requirement to demonstrate aspects of technical knowledge and safety at all stages, up to and even after repository closure. This aspect alone can make monitoring activities necessary and most repository programmes plan for post-closure monitoring.

During the potentially long period prior to repository closure, both future operators and future generations of society will need to make critical decisions about how, when and if to implement various steps in the management of the repository system. A primary objective of monitoring is to provide information to assist in making those decisions. In this context, the key objectives of monitoring of deep disposal systems are seen to be:

1. to provide information for making management decisions in a stepwise programme of repository construction, operation and closure;
2. to strengthen understanding of some aspects of system behaviour used in developing the safety case for the repository and to allow further testing of models of these aspects;
3. to provide information to give society at large the confidence to take decisions on the major stages of the repository development programme and to strengthen confidence, for as long as society requires, that the repository is having no undesirable impacts after closure;
4. to accumulate an environmental database on the repository site and its surroundings that may be of use to future generations of decision makers;
5. to address the requirement to maintain nuclear safeguards, should the repository contain fissile material such as spent fuel or plutonium-rich waste.

In addition to these key objectives, which are all concerned with establishing confidence in the long-term ability of the repository to isolate the wastes properly, monitoring would also be carried out in the operational phase for reasons common to any operating nuclear facility. This would include monitoring to determine any radiological impacts of the operational disposal system on personnel and on the general population, to determine non-radiological impacts on the environment surrounding the repository and to ensure compliance with non-nuclear industrial safety requirements for an underground facility (e.g. dust, gas, noise, radon gas, etc.). None of these operational phase objectives is discussed further here.

In developing a strategy for monitoring, the benefits from gaining data on the behaviour of the system components need to be balanced against any detriments resulting from the process of monitoring. Potential detriments may include:

- degradation of materials in the repository resulting from delay in putting engineered barriers in place whilst monitoring programmes are completed (potentially leading to later imperfect functioning of the engineered barrier);
- hazards arising from keeping a disposal tunnel open for prolonged periods if it is not ventilated continuously, such as the build-up of natural radon or explosive natural gases;
- formation of pathways through the barriers by the installation of monitoring equipment, leading to increased potential for radionuclide migration within or around the repository;

- an increased likelihood of intrusion by people or adverse impacts from dynamic processes on Earth's surface (e.g. flooding) if repository access is kept open to allow monitoring;
- interference with other repository operations;
- radiation doses to personnel carrying out monitoring.

To achieve the five key objectives outlined above, monitoring may be required at various times throughout a repository development programme and at various locations in and around the repository.

A repository development programme involves a series of consecutive phases, beginning with the definition of the disposal concept and moving through site selection and characterisation into the phases of construction, operation and closure. Although there are many ways to "stage" a repository development programme, some typical stages were described in Chapter 4 (Box 3):

- Surface exploration.
- Access construction and underground exploration.
- Construction of the repository.
- Emplacement of waste and near-field engineered barriers.
- Disposal tunnel/vault backfilling.
- Repository backfilling and sealing.
- Post-closure (institutional/non-institutional).

10.2 Monitoring to Establish Baseline Conditions

Site characterisation activities are undertaken to quantify the parameters that will determine repository performance. Some of these parameters may change with time, naturally or due to human activities at the site. Therefore, certain monitoring activities should begin at the earliest possible stage of a repository development programme, before repository construction and operation begins to affect the properties of the site. This early information is important because it allows an understanding to be developed of the nature and properties of the natural, undisturbed environment of the disposal system. So-called "baseline" information is used to underpin each of the first four objectives identified earlier in this chapter. For example, it will be used to evaluate changes that occur in the rock and groundwater system during the construction and operational stages, and, in the post-closure period, to evaluate any impacts that the presence of the repository may have on natural processes and the environment. In practice, the monitoring programme will, therefore, begin during the site investigation programme.

The characteristics of primary interest in the context of establishing baseline information are:

- the groundwater flow field in the host rock formations and in the surrounding geological environment (groundwater pressure distributions, hydraulic gradients, transmissivities, regions of recharge and discharge, etc.);

- geochemical characteristics of groundwaters (redox, salinity, major and trace element concentrations, natural radionuclide content, etc.);
- background levels of natural radioactivity in groundwater, surface waters, air, soils and sediments, animal and plant life;
- meteorological and climatic conditions;
- hydrology of surface water systems, including drainage patterns and infiltration rates;
- ecology of natural habitats and ecosystems.

Baseline data should be established as part of site characterisation activity, from measurements in boreholes and surface investigations. These measurements must be continued for sufficient time to enable the implications of natural (e.g. seasonal) variations, and of any perturbations caused by drilling, sampling and monitoring activities themselves, to be understood. Where important parameter values are found to follow a cyclic trend baseline monitoring will need to be continued until the trend is established with confidence.

10.3 Using Monitoring Information

The construction of a repository will disturb the pre-existing natural system. The subsequent phase of repository operations will cause further changes. Some of these changes may take many years to manifest themselves. Therefore, an important aspect of the monitoring programme will be concerned with changes to the repository environment resulting from mechanical, hydraulic, geochemical and thermal effects. The changes that result from these effects will have particular relevance in regard to the first three objectives defined previously.

10.3.1 Supporting Management Decisions in a Staged Programme of Repository Construction and Operation

Numerous decisions will be required by the operators of the repository and by the regulators during its construction and operational life. Some of these will be made early, in response to site characterisation information, others will be made much later, after all construction has ceased and the repository has been operational for decades. Monitoring information is likely to play a role in the latter. In addition to the major socio-political considerations associated with repository closure, the operators of a repository for long-lived wastes may typically need to consider technical matters such as:

- Adjusting the later stages of repository layout or design in response to long-term monitoring of rock stresses and groundwater flow. For example, long experience of monitoring thermohydraulic effects might lead to changing waste package emplacement spacings and hence heat loading to the rock.

- Modifying waste handling and emplacement procedures, or engineered barrier design or material properties in response to monitoring of the behaviour in and around the initially emplaced wastes.
- Postponing the final design of seals and selection of backfill materials for the various stages of repository closure, basing them on long periods of observing rock stress response, the movement and chemistry of water in excavation damaged zones around openings, or the creep behaviour of plastic formations.
- Deciding when to emplace certain types of buffer material, such as cement, clays, or crushed rock material around ILW packages or concrete vaults, based on monitoring package degradation behaviour, gas production and variations of surface physical and chemical properties of tunnel and vault walls which may affect bonding.
- Choosing the optimum time to backfill and isolate disposal regions completely, in the context of long-term stability of openings.
- Deciding whether to carry out repairs or remedial engineering work to excavation support systems, based on monitoring of rock movements, rock stresses and the degradation of rock bolts, and other support materials.
- Making adjustments to the repository de-watering scheme to account for variable resaturation of different completed regions and consequent long-term changes in the groundwater flow pattern or in groundwater chemistry.

Although the initial plans for a repository will make assumptions about all of these matters as part of the design basis, it is possible that decades of operational experience will allow early decisions to be adjusted and modified to take advantage of what is learned from concurrent monitoring information.

In general, the approach will be to accumulate information from the construction and operation stages to allow the design of the repository to be checked, refined and, where necessary, modified. The aim will be progressively to enhance confidence in the design concept, to allow the successive steps to take place. In some cases, monitoring may be used to provide information about the retrievability of the waste, if such an option is included in the disposal strategy of the country.

This type of monitoring will begin in the early stages of repository construction (e.g. shaft sinking) and continue throughout the construction and operational phases until the repository access ways have been sealed. At each stage, the measurements taken as part of the monitoring programme would be used to provide input to the decision process on whether the necessary confidence exists to take the next step. As appropriate, the monitoring information may also be used to support the necessary submissions to the authorities to seek regulatory approvals.

10.3.2 Strengthening Understanding of System Behaviour

Early in a repository development programme, decisions to accept a site, to go ahead with construction and to emplace waste in the repository will have been based in part on the results of performance and safety assessments that will have evaluated the long-term, post-closure behaviour of the disposal system. Much of the information to underpin these assessments will have been derived from site characterisation work and supporting R&D, often carried out over many years. In order to proceed with these early steps, sufficient confidence will need to have been accrued in the ability of the conceptual models and databases used in the assessment studies, to represent future system behaviour adequately.

Although endorsement of the early programme steps must be based on having sufficient confidence in post-closure safety, it is clear that the opportunity will exist to test and strengthen understanding of some aspects of system behaviour further during the long pre-closure period, possibly several decades. It is possible that regulators, decision makers or society in general may require this extra time to develop the confidence that its early decisions to place waste under the ground were justified, and that the final step, that of repository closure, can go ahead. It is not possible today to say how long this time might be; it is largely a societal decision and may vary from country to country. In some programmes, it might be intended to retain the option of waste retrievability during the whole period.

The aspects of a safety case that can be tested further on the basis of protracted monitoring during the pre-closure period include understanding of:

- the groundwater flow field, which should respond in a broadly predictable fashion, over many years, to excavation and repository operations, in which some regions of the rock may at times be drained and later partially isolated and left to resaturate;
- groundwater chemistry, which may change in composition as different regions of the rock respond to underground operations, thus allowing further tests of hydrogeological and hydrochemical models of the host environment;
- the hydraulic and mechanical behaviour of important structures in the rock;
- the thermal field around repository structures (emplaced heat-emitting wastes, large concrete vaults);
- the response of underground structures and the groundwater system to seismic events (more important in some regions of the world than others);
- the resaturation behaviour of regions of a repository that have been partially completed and isolated from operational areas, particularly the uptake of water by buffer materials whose properties are an important aspect of long-term performance in some concepts;
- chemical interactions between engineered barriers and the rock/groundwater system.

Each of these topics is likely to have been considered, and predictions made about their impacts on performance, when presenting a safety case for waste emplacement. An additional twenty, forty or more years of observations can serve to improve understanding. The intensity of monitoring activities will be decided by weighing against one another the types and resolutions of data required, the need to avoid negative impacts on safety and the costs of monitoring. One approach to monitoring to assist with major decisions may be to identify a small representative part of the repository (or parts, to reflect variable rock conditions) as a demonstration region. This area could be densely instrumented to monitor these factors, with much sparser (or no) instrumentation or observations being located elsewhere. This approach has been proposed in Sweden where pilot repository operation with a small fraction of the total inventory is foreseen and in Switzerland where it has been proposed to have separate test, pilot and disposal caverns (EKRA, 2000).

It has to be accepted that some aspects of system behaviour will most likely differ from those envisaged at the outset. To maintain and improve confidence towards the final step of closure, it will be necessary to show convincingly that these differences do not affect long-term performance, and that the safety case is robust. It also has to be accepted that, if the reverse proves to be the case, then a different disposal solution should be sought. This perception lies behind the demand for maintenance of ease of retrievability in some programmes (see Chapter 4).

In the context of the above discussion, it should be noted that the monitoring period is extremely short compared with the period of expected isolation. Also, the relevance of monitored parameters to long-term safety and the allowable deviations from expected values before action levels are reached are very difficult to quantify. So far, it is thought that no programme has yet developed a comprehensive monitoring concept which takes these issues into account.

10.3.3 Input of Monitoring Data to Societal Decision Making

Because of the high public profile of waste disposal programmes, there are several critical points in a repository development programme that are likely to demand input from a broader range of stakeholders than the repository operators and regulators alone. It might, for example, be expected that society, having endorsed the solution of geological disposal at a political level, would want to be involved in making decisions on at least the location of a repository, and on approving its eventual backfilling, sealing and closure. In fact, there is an increasing tendency to develop repositories in a stepwise fashion, with the decision to proceed (or not) to the next step being based on a transparent evaluation of lessons learned during previous steps (NRC, 2001a). Monitoring data will obviously serve as input to many of these steps.

The siting decision is unlikely to require any specific input from monitoring programmes, other than those used to establish baseline information for the sites under consideration. It will be based on technical information, site characterisation data, safety assessments and a range of other technical and non-technical factors. A decision to backfill and seal a repository is, on the other hand, likely to demand a close evaluation of monitoring information collected over what could, by this stage, be many decades of repository operation. For long-lived waste repositories in general, several technical factors derived from monitoring programmes will have to be taken into account, including the structural stability of repository openings and roof support systems, the developing properties of engineered barriers, interactions between engineered barriers and the rock–groundwater system, and the degradation state of waste packages, particularly if not surrounded by engineered barriers. For the specific case of HLW, the latter aspects are clearly of much less significance than for a concrete vault repository for LL-ILW.

Society will also have to decide on the level of post-closure control that it wishes to see exercised over the repository. Some will see no need for institutional control period on the land overlying and surrounding the repository location once it has been sealed and the site decommissioned. Others may wish to see that access and activities on the site are limited to varying extents. In either case, it is most probable that there will be a demand to have the environment of the repository monitored. The period over which this monitoring might continue is impossible to predict, as it will depend entirely on the views of society in many decades time. It can only be assumed that monitoring would continue until all concerned parties are completely comfortable with the post-closure status of the repository, and confident to suspend or wind-down monitoring programmes.

A key issue in post-closure monitoring is the need for caution, such that any programme of measurements and instrumentation is not intrusive into the isolation barriers, with possible detrimental impacts on long-term performance. Monitoring that makes use of remote sensing or periodic surface sampling is to be preferred. Such a programme may focus on detecting ground movement and heat flow (perturbations to any regional patterns which might be caused by the repository), as well as local water quality.

10.3.4 Accumulating an Environmental Database

Much information will accrue during the lifetime of a monitoring programme for an operational repository. A large amount of this information will have been gathered and been used specifically to meet the previous three objectives, and will then have limited direct use after repository closure, although it should be maintained as part of the record of the disposal activities.

We are not in a position today to be able to judge conclusively what types of information future generations may find useful after the repository has been closed. They may want to be able to take decisions about future land use, the context of which cannot presently be envisaged. Post-closure monitoring of some key environmental indicators could be required by society for some indefinite period of time. In order to put these post-closure measurements into context, it would be valuable to be able to pass a comprehensive database of long-term environmental observations along to future generations of decision makers. Such a database might be expected to contain long-term time series observations (collected over, possibly, tens to hundreds of years).

10.3.5 Nuclear Safeguards

As discussed in Chapter 2, if the repository accepts waste with significant quantities of fissile radionuclides, such as spent fuel or plutonium-rich waste, nuclear safeguards are likely to be an important issue (Pellaud and McCombie, 2000). Even after repository closure, it may not be possible to declare the waste "practically irrecoverable" and to meet the requirements for termination of safeguards (Fattah, 2000). Thus, some kind of control of the site might be required. Any safeguard-related monitoring would be aimed at providing assurance that no unlawful retrieval of material from the repository takes place. Some essential requirements of a safeguard-related monitoring programme are that:

- monitoring programme should not reduce the safety of the disposal system; therefore intrusive methods relying on the emplacement of instrumentation in the isolation barriers would not be acceptable;
- since the requirement for safeguard-related monitoring can be expected to last for a long time, the efficiency and cost-effectiveness of the system should be maximised.

In practice, it may be possible to ensure that no drilling or mining activity, which would be a prerequisite for the retrieval of any nuclear material, is taking place at the repository site, by relying on periodic site observations by means of aerial photography or satellite imagery. If tighter surveillance of the site were required, this might be achievable through a network of microseismic stations designed to reveal the emission of energy associated with drilling and mining (Peterson, 1998). As noted in the introduction, this area will clearly require further consideration as repository development programmes develop.

10.4 Post-Closure Monitoring: Problematic Issues

The justification for monitoring before and throughout the repository operational phase, and the methods that can be employed to do this, are relatively straightforward. This is because the objectives (as outlined above) are clear, the techniques are feasible (with continued access giving the ability to maintain and replace devices), and the timescales (although perhaps long) are definable. For post-closure monitoring, the challenges are greater.

The expectation must be that no direct releases will be measured over monitoring times that are very long (hundreds of years) compared to any normal activity — otherwise there is no way that the repository could be licensed and built. Indirect efforts on the rocks, the engineered barriers, the hydrology, etc. can certainly be measured, but there are real problems relating most of the observable parameters to long-term repository safety (on the timescales of many thousands of years). Measurements techniques should not be of an intrusive type that requires penetration of the sealed repository safety barriers; remote sensing tools would need repair, recalibration and replacement. Nevertheless, post-closure monitoring is envisioned in most programmes — even if the expectation is of zero impact measurements, this can help build public confidence over decades.

To design a post-closure monitoring programme that is scientifically sound and also practicable is a challenging task that has nowhere yet been fully completed. Any plan for measurements meant to confirm safety or provide public reassurance should in principle:

- identify the parameters to be measured and show their relevance to system performance;
- identify the means by which the monitoring shall be done, and show that it is possible to get a meaningful signal;
- select the point (or points) at which the measurements are to be made and whether they are to be made in a remote or in an intrusive way;
- show how the signal is going to be interpreted;
- present the range of normal readings and its time dependence (taking background levels and fluctuation into account);
- discuss possible deviations in the normal readings and their significance for the safety case;
- present the time duration of the monitoring;
- identify safety-motivated action levels, and specify the actions to be taken;
- evaluate the reliability of the monitoring system, including possible actions required if malfunctions occur;
- document the plan.

It is difficult or impossible to develop a programme fulfilling all of these objectives even if a large effort were to be invested. This is one reason for the early scepticism of many technical communities concerning the value of post-closure monitoring when the safety case has been already made.

Recently, particularly in the USA where the Yucca Mountain repository project is moving into an implementation phase, there has been increasing interest in monitoring. The terminology often used in the USA is a wider term "performance confirmation" that includes all of the activities concerned with examining whether a repository system is behaving as was expected. A recent workshop (EPRI, 2001) looked at all of the approaches and techniques available.

The participants at the workshop also developed an interesting list of what they call "traps" awaiting those developing performance confirmation approaches. The adapted list, with added comments is as follows

- Do not agree to things that cannot be done. Often the concerned public at or near a site can request monitoring measures that are not feasible, technically or economically. This should be openly acknowledged and explained.
- Do not agree to measure parameters that do not affect performance. This only adds confusion and expense.
- Do not claim safety based on monitoring of too limited duration. It must be acknowledged that any feasible monitoring period will be trivially short compared with the waste hazardous lifetime.

- Do not require unnecessary accuracy or precision in measurements. This is a common way to incur unnecessary expenses.
- Make sure the target parameters can indeed be measured in the given environment. In or close to the waste packages, high radiation doses can prevent some measurements.
- Keep something in the programme to satisfy periodical interests. Being too hard and refusing all suggestions from stakeholders, especially locals, is counter-productive.
- Do not assign excessive levels of conservatism on any subsystem just because it is in any case easy to meet the requirement. Levels of conservatism should reflect the levels of scientific confidence.
- Do not forget that the implementor has decades in which to do the work. This gives time for new developments and also makes too exact specifications premature.
- On the other hand, do not unnecessarily postpone with current technical concerns that can be dealt with now.
- Never forget that technical approaches cannot succeed unless institutional aspects are considered also in the process.

The sensible caveats listed above, together with the challenging requirements on a monitoring programme defined earlier, will hopefully allow the issue to be addressed by a sensible approach capable of consensual agreement. The consensus that must be reached is between the concerned public (who are often after an unachievable demonstration of perfect safety), the scientific community (the most extreme of whom advocate collection of almost all measurable data), the safety analyst and regulator (who must have any key safety data) and the implementor (who must finance a possibly very long-lasting measurement campaign).

10.5 Conclusions

It is widely accepted that the long-term safety of geological disposal should not rely on a continued capability to monitor a repository after it has been sealed and closed. Ethically, future generations should have no monitoring burden placed upon them. Nevertheless, they may wish to monitor. It is thus the responsibility of repository implementors to consider how an appropriate monitoring system might be designed, without compromising safety. However, there are several more immediate applications of monitoring information which the repository designers and operators should be encouraged to consider.

Monitoring a deep geological repository and its environment would be carried out principally as an aid to decision making. Some of the decisions are technical and can be based on specific monitoring parameters. Some are societal and will depend on monitoring programmes having enhanced public confidence in the system to a sufficient extent. There is a variety of objectives for carrying out monitoring, and a range of applications for monitoring information. Some of these objectives require monitoring to be carried out throughout the operational life of a repository, and the majority of them also require the clear establishment of a set of baseline information related to the largely undisturbed condition of the site.

The main types of decision that will be supported by using monitoring data in the period up to repository closure, are those concerned with moving from one operational stage to another in the management of the facility. These may be small steps, such as decisions on the eventual inventory and backfilling for a specific section of the repository. Some of these decisions, particularly the decision to close a repository, have wider significance, and may need to be taken not just by the regulator but by means of a consultative process involving society at large.

At the point of closure, it becomes considerably more difficult to reverse disposal and to retrieve the wastes. Consequently, at this time, sufficient confidence in the capability of the disposal system to contain the wastes safely must have been achieved. Thus, a further aspect of monitoring during operations is to enhance understanding of those aspects of the underlying safety case that it is feasible to address over a period of several decades.

Up to this point, simple, non-prescriptive regulations could be devised that merely require the implementors to mount a programme of monitoring that will be adequate for making decisions at each step on the move towards facility closure. Since the implementors will need to provide details of operational experience and safety analyses of future performance at each step, it would be expected that regulations should require a comprehensive programme of monitoring to exist in support. The regulators role would be to review the proposals for monitoring, to receive all monitoring data and to draw their own conclusions based on their independent analysis of the information provided.

After closure of the repository, there may be a continued demand for monitoring of the site and the surroundings, even if there is no control over land use. The needs of future generations can be helped if a comprehensive and continuous, time series environmental monitoring database is passed on to them. However, there is no reason why regulations should *require* that monitoring continues after institutional control ceases: indeed, agreement for the cessation of institutional control would be based in part on the provision of "zero radiological impact" returns from monitoring programmes that may have been in place since before repository construction began. There are also potential problems in that proponents of monitoring programmes may claim the ability to carry out actions that are not really practicable over long timescales, or may also imply that monitoring — even for some hundreds of years — is likely to bring direct safety-relevant data. The "traps" listed earlier from the EPRI workshop should be heeded by more of those developing monitoring programmes.

The first of the above problems raised an important practical point that has not been directly addressed in this chapter. It concerns the problems of developing and operating instrumentation that will remain sufficiently reliable for use over the potentially long monitoring periods in relatively hostile environments. Although surveillance and monitoring for nuclear safeguards purposes appears feasible using available methodology, both during and after repository operations, further enhancements in the robustness of instrumentation may be needed. In common with any nuclear plant, it requires the co-operation of the operators. At some time after closure, responsibilities will be transferred to the national government or to an international safeguards regime.

Chapter 11

Preserving Records of the Existence of a Repository

The least predictable of future scenarios that could lead to exposure to radioactivity are those in which people, perhaps in the far future, interfere with or intrude into the repository, either accidentally or intentionally, as discussed in Chapter 9. The probability and consequences of such intrusion can be influenced by the amount of information that future societies may have about the wastes. Chapter 9 recognised that human intrusion is difficult or impossible to exclude and discussed measures for reducing potential impacts. There are, nevertheless, steps which can be taken to reduce the probability of human intrusion by means of preserving information about the repository for future generations. The present chapter addresses those measures which are intended to remain effective after the period of active control and monitoring, described in Chapter 10, has ended. The issue of assembling and preserving records on the waste inventory out to these very long times is of great importance to repository programmes.

The key questions which are of interest are:

- Why, and for whom should records be preserved?
- For how long should, and can, record-keeping systems function?
- How, and where, can records be reliably preserved?
- What research has been done in this area and what remains to be done?

11.1 Historical Overview

Obviously both deliberate and inadvertent intrusions can be prevented for as long as active control is maintained over the repository site. Restricted access, fences, guards and other security measures can all form part of an active institutional control system. Such controls are assumed to be capable of being implemented and maintained for relatively long periods of 100–300 or even 500 years. They form part of the overall safety system for near-surface repositories for short-level wastes in, for example, France and Spain. In the USA, some of the contaminated sites from earlier defence activities are acknowledged to be incapable of full remediation, so that they have been said to require indefinite active institutional control (NRC, 2000b).

For high-level wastes and spent fuel, the timescales for which the radiotoxicity remains high are long, extending to tens or hundreds of thousands of years. For such wastes, the US National Research Council (NRC, 1995) has concluded in agreement with others that "*it is not reasonable to assume that a system for post-closure oversight of the repository can be developed, based on active institutional controls, that will prevent an unreasonable risk of breaching the repository's engineered barriers or increase the exposure of individual members of the public to radiation beyond allowable limits*". This means that the repository safety system should be based on passive engineered and geological barriers and that long-term reduction of intrusion risks should be based on what the EPA calls "passive institutional control" (PIC) measures (see Box 8), to distinguish them from active measures (AIC).

It is clear from Box 8 that the only real measure of "control" in this definition of PIC is land ownership and regulations regarding land use, and this only acts as a control if there is some effective operational system of controls on land use planning. This rather challenges the concept of PIC and indicates that, once active control has ceased, the only remaining protective measure relies on people having access to records and markers, and taking sensible decisions after using them.

Given the scenarios that could lead to humans coming into contact with the wastes, is it more prudent to ensure that information on the repository site and contents is widely and openly available, or would it be better to have an unmarked site and restricted information (as discussed in Chapter 9)? Although early

Box 8: USEPA Definitions of Active and Passive Institutional Control

Passive institutional control (PIC) means: (1) Permanent markers placed at a disposal site, (2) public records and archives, (3) government ownership and regulations regarding land or resource use, and (4) other methods of preserving knowledge about the location, design, and contents of a disposal system.

Active institutional control (AIC) means: (1) Controlling access to a disposal site by any means other than passive institutional controls; (2) performing maintenance operations or remedial actions at a site, (3) controlling or cleaning up releases from a site, or (4) monitoring parameters related to disposal system performance.

(*from USEPA Regulation 40 CFR 191*)

suggestions were made on keeping information restricted (Rochlin, 1977), the current consensus (see for example IAEA, 1999c) is that measures should be taken to ensure that all relevant information on a deep geological repository is assembled and preserved for future generations. Already, in its regulations 10 CFR 60 and 40 CFR 191 (USNRC, 1981; USEPA, 1993), the United States required that a repository site be marked with the most permanent markers practicable. Later legislation on Yucca Mountain (USNRC, 2001; and EPA, 2001c) also requires measures to preserve information on the site.

The most extensive studies on the issues of markers and records have, in fact, been done in the USA. Early work was done in the 1980s under contract to the Office of Nuclear Waste Isolation (ONWI) on markers (Kaplan, 1982, 1986; Berry, 1983) and on communications (Sebeok, 1984; Tannenbaum, 1984; Weitzberg, 1982). At a later date, the challenges of developing marker systems and defining information for archiving were tackled by the WIPP programme in the course of its Compliance Certification Application (CCA), published in 1996 (DOE, 1996). This document, which will be discussed in more detail later in this chapter, contains the most complete set of proposals documented to date for marking a geological repository and for archiving appropriate data.

The next most comprehensive studies were those performed in Scandinavia in the mid-90s. These concentrate on the aspect of information conservation and retrieval. The main results are provided by Eng et al. (1996).

Other countries have done comparatively little work on markers or on record keeping. However, concepts have been developed in also France, Switzerland, Spain and Canada.

11.2 Rationale for Maintaining Information on a Repository

The basic objectives of passing on information as formulated in IAEA (1999c) is to ensure that, even for future unpredictable natural disasters or human actions, "*the likelihood of adverse health consequences from such events [will] be reduced by providing information and warning to future generations regarding the presence of the waste and its potential hazard*". Thus, as long as the presence of the repository is known, the probability of inadvertent intrusions is reduced and, in the case of deliberate intrusion, sufficient information on the repository inventory and design will be available to allow planning of the actions in such a way as to avoid unacceptable doses to the workers involved.

These objectives give guidance on the form and the content of the information to be preserved and passed on via a system of passive institutional controls. There are two basic approaches. One is *direct communication* to future populations, by leaving at the site durable markers with appropriate information. The other is indirect, in that information is stored and archived at a diversity of off-site locations, with appropriate measures for maintenance being taken by successive generations.

Methods must be developed to ensure that the records are preserved and that society remains aware of their existence and retains the capability to interpret the records.

In the following text, individual sections are devoted to the principles involved in:

- designing markers or monuments to be placed at the repository sites;
- implementing systems for creating and maintaining off-site records.

11.3 Monuments and Markers at Repository Sites

11.3.1 Design Principles

It was pointed out in Chapter 9 that, although marking a site is intended to discourage human actions, history has shown that monuments (e.g. pyramids, stone circles, ancient earthworks) actually tend to attract intruders. The common driving force is the prospect of valuable finds and this may be counteracted for a repository by information showing that the contents are dangerous. Another driver, however, is simple curiosity and, should the information on hazards be lost or not interpretable, then monuments could well increase the probability of the "naïve intrusion" scenario introduced in Chapter 9. An intriguing response to this, proposed in all seriousness in a French study (Raimbault, 1993), is to erect the monument at some distance (20–30 km) from the site and include on it only encoded information on the actual location.

Most work, however, is aimed at developing durable, comprehensible monuments and markers at the repository site. Based on studies of archaeological objects Kaplan (1986) and Adams and Kaplan (1986) developed a list of design criteria for effective markers and monuments. These include the following:

- markers should delimit the area overlying a repository, rather than a single point: this is best done by using multiple markers and ensuring that each marker is within sight of at least one other;
- the materials of which the markers are composed should be natural of little intrinsic value and durable (e.g. granite, basalt, clay);
- the monuments should be large enough to avoid removal (several metres);
- redundancy is valuable in case single markers are, nevertheless, removed or shifted;
- messages should be engraved in large letters on surfaces oriented and positioned to minimise erosion;
- the messages on markers should use a combination of symbols (pictograms) and language(s): each can help the other to be interpreted and the (more complex) language messages are needed to convey sufficient information;
- markers should be both on the surface and sub-surface (at depths sufficient to deter souvenir hunters).

11.3.2 Types of Markers

Massive surface markers can be used to delineate the area underlain by the repository. These can be in the form of a defined pattern of monuments or a larger area of ramparts or earthworks (see Fig. 11.1). Archaeological analogues of both types indicate that they can be very durable. The markers should bear sufficient information to warn future societies of the hidden hazard and should indicate locations where more detailed information is available. Some of the markers should be underground in case the surface monuments are lost. Underground monuments can be elaborate edifices similar to the surface structures, bearing the same level of information. They can also be simpler distributed markers spread around at depth in a random fashion (to make deliberate collection by souvenir hunters less easy) but densely enough to raise the probability of any future exploration programme at the site actually encountering at least one.

11.3.3 Information Content of Monuments and Markers

The content of the information should, at a minimum, allow future societies to recognise that there is a hazard to humans and to be able to locate this hazard. A combination of symbols and a variety of world languages is recommended to increase the probability of messages being comprehensible in the far future. Specific proposals have been made Adams and Kaplan (1986), Kaplan (1986), Tannenbaum (1984) and others. These range from complex combinations of symbols to the simple common trefoil used today to indicate radiation hazards.

The early US work on markers and monuments was followed up in the 1990s in preparation for developing specific designs for WIPP, the first purpose-built deep geological repository (Benford et al., 1991; Nolin, 1993; Trauth, 1993). The designs which were then chosen for the 1996 WIPP compliance certification (DOE, 1996) are described in more detail below.

11.4 Record Keeping/Archiving in Disposal Programmes

Archiving information should not be regarded as an alternative to markers and monuments but as a complementary strategy which increases redundancy and provides more detailed information. Some of the earliest and most thorough work in this area was done in the scope of a Nordic countries' project (Eng et al., 1996). This was based in part on historical studies of the fate of various ancient archives such as those maintained by the Vatican, the oldest records of which date back to the end of the 12th century (Paztor and Hora, 1994).

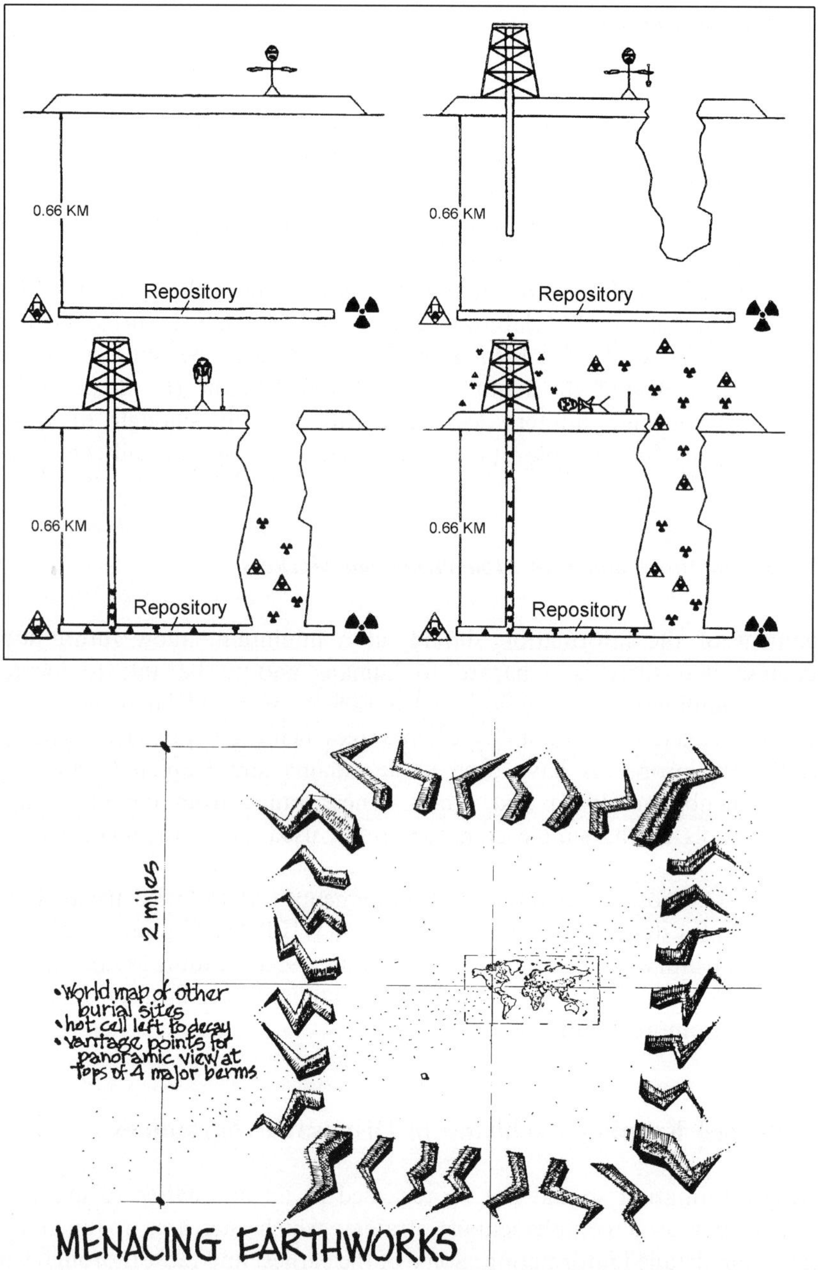

Fig. 11.1. Concepts for marking the site of the operational WIPP repository in New Mexico, USA. Top: a pictograph communicating the idea of radiological hazard if the repository is disturbed. Bottom: massive earthworks on the surface to define the boundaries of the repository zone in a manner thought to convey menace.

11.4.1 Contents of Archives

The information required in an archive is a function of the objectives of those assumed to be future users of the information. To avoid inadvertent hazards to those carrying out a future exploration programme, one need only know the location of the repository and that the contents represent a real hazard. Future groups wishing to retrieve the disposed materials or to carry out remedial actions on the repository will need more information.

The Nordic project proposed a hierarchical information system with seven levels (Eng et al., 1996) (Table 11.1).

Levels 6, 5 and 4 correspond to what the study called Primary, Second Level and Third Level information sets. It was proposed that these should be archived at, respectively, the site, regional archives, and multiple national and/or international archives.

In a similar fashion, those working on the WIPP project defined Message Levels to be preserved at different locations (Table 11.2).

Table 11.1

Level 1	Rudimentary hazard warning via a sign or symbol
Level 2	A one sentence warning message
Level 3	Basic information such as a waste inventory description
Level 4	Detailed repository information (~100 pages of scientific reporting)
Level 5	The complete technical records on the repository
Level 6	Much broader information including the above but also documentation on legislative and societal issues
Level 7	All information available in society as the existence and features of the repository

Table 11.2

WIPP message level	Characteristics (locations)
I	The site is man-made (should be obvious at the site from the form of structures and the effort involved)
II	Something is dangerous (on monuments or markers)
III	What, why, when, where, who, how (detailed information on monuments)
IV	Complex information in multiple languages (in surface and buried documents rooms)
V	Complete rule making (in distributed archives)

More specifically, the information about the repository which it is suggested requires preservation includes:

- the geographical location (coordinates and depth);
- the radioactive waste inventory;
- the complete design of the repository (including details of all engineered barriers);
- individual waste package contents and locations;
- information on when, and by whom, the repository was operated;
- all input data for, and results of, the final safety analyses;
- details of repository operation, closure and sealing;
- environmental and ecological data collected throughout the repository programme.

11.4.2 Record Management System (RMS)

A record management system (RMS) must be implemented at an early stage in repository development in order to ensure that the appropriate data are collected, maintained, and transferred to appropriate archives. The hierarchical system proposed by the IAEA (IAEA, 1999c) is based on three categories of information:

- *Primary level information (PLI)*: this is the complete set of records gathered on the repository through its history.
- *Intermediate level information (ILI)*: a more condensed set, mainly consisting of information needed to meet regulatory and legislative requirements (and indicating where the PLI can be found).
- *High level information (HLI)*: more condensed information needed by future generations to understand the repository; to be archived nationally and internationally.

The IAEA gives a table of HLI records, expanding in more detail the above list prepared by WIPP and adding two significant points, namely a description of the RMS itself and data needed to meet nuclear safeguards requirements.

Of equal importance to the records definition is the specification of appropriate procedures for implementing an RMS. The tasks involved in establishing an RMS are

- define the exact information to be saved;
- define the physical form for the records, and their locations and periods of retention throughout the system;
- allocate responsibilities for collection and maintenance of records;
- define measures to ensure that the information will remain accessible and comprehensible;
- consider remedial actions in the event of the deterioration of records;

- coordinate with archivists at regional, national and international levels to ensure adequate distribution of records;
- specify an overall system quality assurance programme to ensure that all steps are properly carried out.

11.4.3 Physical Forms of Information Records

As is pointed out by the IAEA (1999c), there are some key requirements which must be met by any medium used to store the repository information.
These include:

- sufficient capacity for storage of information at the particular level in question;
- physical and chemical stability in order that legibility is preserved for a long time;
- easily copied or transferred to another medium without significant loss of information;
- retrievable for long periods;
- readable and understandable;
- capable of being protected from unauthorised amendments;
- not comprised of materials with a high intrinsic value.

The various media options are discussed in IAEA (1999c) using information extracted from the ISO 9706 standard (ISO, 1994). Before reviewing conventional archive materials, it is worth noting that the markers and monuments described above are already a primitive sort of archive, especially the proposed US schemes for duplicated vaults with information engraved on durable natural stones. These stone tablets fulfil most of the above requirements, with the important exception of the first one.

For archived data in the normal sense, important questions are the storage capacity of the medium and the resistance of the records to ageing. An important distinction in this respect is between two types of ageing that can both lead to problems: physical ageing (e.g. fading of inks, loss of magnetism in tapes or discs) and logical ageing, in which the technology for reading data can be lost (e.g. punch card readers, outdated computer software). The modern media considered by the IAEA are as follows in Table 11.3.

Relative to the vast data quantities gathered, stored and archived for innumerable other technological reasons today, the amount of information required to be kept for repositories is very minor. Even the most detailed description of a repository project can be easily stored on, for example, a few compact digital discs. The extremely extensive reports on the Compliance Certification for the WIPP repository in the USA were distributed widely as a set of four compact discs (DOE, 1996). Data of this type are easily and cheaply copied and can thus be archived at numerous places in order to ensure redundancy. Accordingly, for the immediate future, there

Table 11.3

Medium	Key characteristics
regular paper	lifetime a few decades; readable without special tools; copies easily; bulky
permanent paper	lifetime several hundred years (under controlled conditions); otherwise as above
microfilm	expected lifetime 100–200 years; larger capacity; readable with simple tools; copying or transferral to other media decreases quality
magnetic disc	lifetime 5–10 years; very large storage capacity; high maintenance requirements; needs special hardware and software; difficult to detect unauthorised changes; data can be destroyed by magnetic fields
optical disc	expected lifetime over 100 years (but very uncertain); otherwise as for magnetic disc
future options	modern data storage and processing technology is continually working towards devices with increased storage capacity; this is an area in which there will certainly be developments in the state-of-the-art during the decades in which repositories are constructed and operated.

seems to be no incentive to seek definitive solutions for data storage media. Especially as any technology identified today will undoubtedly be superseded with the lifetime of a repository.

The important challenge is to establish a structured organisational system which ensures that data collection begins early in the repository project, that the data are quality assured and can be altered only with proper authorisation, and that procedures are established for maintaining the records in a flexible easily readable and easily copied form.

11.5 The Example of the USA

The early concepts developed in the WIPP programme in New Mexico are now being implemented and also being adapted to the Yucca Mountain project in Nevada. Other countries, most notably Sweden, have also tackled the issue of passive controls, but their work is heavily based on the US studies. Accordingly the USA studies are described here in more detail.

11.5.1 Regulatory Requirements

For disposal of HLW, spent fuel and long-lived wastes, the USA has a comprehensive set of legislation from EPA and NRC. This also covers requirements concerning PIC. The regulation 10C FR 60 contains explicit requirements, as detailed in Box 9.

The most recent and most relevant requirements are contained in the EPA rules for demonstrating compliance at WIPP; these are reproduced in Box 10.

Box 9: Requirements for Record Keeping in 10 CFR PART 60 — Disposal of High-Level Radioactive Wastes in Geologic Repositories

§60.51 License amendment for permanent closure.

(a) DOE shall submit an application to amend the license prior to permanent closure. The submission shall consist of an update of the license application submitted under §§60.21 and 60.22, including:

(1) A description of the program for post-permanent closure monitoring of the geologic repository.

(2) A detailed description of the measures to be employed — such as land use controls, construction of monuments, and preservation of records — to regulate or prevent activities that could impair the long-term isolation of emplaced waste within the geologic repository and to assure that relevant information will be preserved for the use of future generations. As a minimum, such measures shall include:

(i) Identification of the post-closure controlled area and geologic repository operations area by monuments that have been designed, fabricated, and emplaced to be as permanent as is practicable; and

(ii) Placement of records in the archives and land record systems of local State, and Federal government agencies, and archives elsewhere in the world, that would be likely to be consulted by potential human intruders — such records to identify the location of the geologic repository operations area, including the underground facility, bore-holes and shafts, and the boundaries of the postclosure controlled area, and the nature and hazard of the waste.

(3) Geologic, geophysical, geochemical, hydrologic, and other site data that are obtained during the operational period pertinent to the long-term isolation of emplaced radioactive wastes.

(4) The results of tests, experiments, and any other analyses relating to backfill of excavated areas, shaft sealing, waste interaction with the host rock, and any other tests, experiments, or analyses pertinent to the long-term isolation of emplaced wastes within the geologic repository.

(5) Any substantial revision of plans for permanent closure.

(6) Other information bearing upon permanent closure that was not available at the time a license was issued.

(b) If necessary, so as to take into account the environmental impact of any substantial changes in the permanent closure activities proposed to be carried out or any significant new information regarding the environmental impacts of such closure, DOE shall also supplement its environmental impact statement and submit such statement, as supplemented, with the application for license amendment.

60.71 Records and reports.

(a) DOE shall maintain such records and make such reports in connection with the licensed activity as may be required by the conditions of the license or by rules, regulations, and orders of the Commission as authorised by the Atomic Energy Act and the Energy Reorganisation Act.
(b) Records of the receipt, handling, and disposition of radioactive waste at a geologic repository operations area shall contain sufficient information to provide a complete history of the movement of the waste from the shipper through all phases of storage and disposal. DOE shall retain these records in a manner that ensures their usability for future generations in accordance with §60.51(a)(2).

Box 10: PIC Requirements in EPA regulation 40 CFR 143

§194.43 Passive Institutional Controls.

(a) Any compliance application shall include detailed descriptions of the measures that will be employed to preserve knowledge about the location, design, and contents of the disposal system. Such measures shall include:
(1) Identification of the controlled area by markers that have been designed, and will be fabricated and emplaced to be as permanent as practicable;
(2) Placement of records in the archives and land record systems of local, State, and Federal governments, and international archives, that would likely be consulted by individuals in search of unexploited resources. Such records shall identify:
(i) The location of the controlled area and the disposal system;
(ii) The design of the disposal system;
(iii) The nature and hazard of the waste;
(iv) Geologic, geochemical, hydrologic, and other site data pertinent to the containment of waste in the disposal system, or the location of such information; and

(v) The results of tests, experiments, and other analyses relating to backfill of excavated areas, shaft sealing, waste interaction with the disposal system, and other tests, experiments, or analyses pertinent to the containment of waste in the disposal system, or the location of such information.

(3) Other passive institutional controls practicable to indicate the dangers of the waste and its location.

(b) Any compliance application shall include the period of time passive institutional controls are expected to endure and be understood.

(c) The Administrator may allow the Department to assume passive institutional control credit, in the form of reduced likelihood of human intrusion, if the Department demonstrates in the compliance application that such credit is justified because the passive institutional controls are expected to endure and be understood by potential intruders for the time period approved by the Administrator. Such credit, or a smaller credit as determined by the Administrator, cannot be used for more than several hundred years and may decrease over time. In no case, however, shall passive institutional controls be assumed to eliminate the likelihood of human intrusion entirely.

11.5.2 Measures Proposed for Implementation at WIPP

In response to the requirements set by the USEPA, the WIPP project developed an extremely detailed concept both for markers and monuments, and for data archiving. This is the most comprehensive set of proposals to date (DOE, 1996).

DOE has interpreted the regulatory language of EPA as a mandate to develop and implement a system of passive institutional controls consistent with the components listed in the EPA's definition. The intention is to protect the integrity of the disposal system for as long as practicable after disposal. Three subject areas must be addressed:

(a) detailed descriptions of the passive institutional controls must be provided;
(b) the period of time that the passive institutional controls are expected to endure and be understood must be estimated;
(c) credit for the passive institutional controls in reducing the likelihood of inadvertent human intrusion in performance assessments must be justified for the proposed time period.

The monument and marker components of the WIPP PIC system consist of:

- *Monuments that define the boundary of the withdrawal area:* These are large monuments erected on the surface at the controlled area boundaries. To facilitate fabrication and shipping of the monuments, each will consist of two separate stones connected by a tendon joint. The large monuments will be engraved with Level II and III messages and Level IV pictographs, using the terminology defined earlier.

- Markers at the footprint of the repository that consist of granite monuments that identify the outer boundary of the subsurface facility: The monuments intended for marking the repository footprint will differ from those marking the controlled area boundaries. Each footprint monument will be inscribed with the Level II and III messages in seven languages, the six official United Nations languages (English, French, Spanish, Chinese, Russian and Arabic) and Navajo.
- *A berm surrounding the repository footprint:* This should be an enduring structure, sufficiently massive to provide reasonable expectation that it will endure for thousands of years. The construction materials should be reasonably available to the WIPP site and have little intrinsic value. To provide a distinctive magnetic signature for the berm, large permanent magnets will be buried at intervals in the berm. Similarly, to provide a distinctive radar-reflective signature unique from the surrounding terrain, trihedrals fabricated from metal will be buried in the berm.
- *A buried room halfway between the information centre and the berm and a buried room halfway between the berm and the hot cell:* The buried storage rooms contain Level IV messages and associated diagrams engraved on the walls. The rooms will be made of granite with a minimum number of joints. Individual walls, the floors, and the roofs will comprise single granite slabs joined only at the edges.
- *An information centre on the surface at the centre of the repository footprint:* In addition to the buried storage rooms, an information centre will be located on the surface providing access to the same information that is contained in the buried rooms. Details regarding the location of one of the buried storage rooms, and identical information, will be contained in the information centre.
- *Randomly spaced buried markers distributed across the repository footprint:* These small warning markers have Level II messages in the seven languages previously listed. However, each marker will have the message in only one of the seven languages. The warning markers will be made of a diversity of durable materials, such as granite, aluminium oxide, and fired clay, thus improving the likelihood that at least some of the markers will endure for thousands of years.

The record-keeping activities defined for WIPP comprise the following:

- sets of records distributed to national and international archives;
- sets of records distributed to record centres locally, nationally, and internationally (both those of a general nature and those specialising in land and resource use);
- government control and land-use restrictions;

- other means of communication, such as encyclopaedias, dictionaries, textbooks, and various maps and road atlases.

The archived material will include information that defines the location, design, content, and hazards associated with the WIPP. The amount of information will be more extensive than that available within the permanent marker system at the repository location. Specific documents in the archived information portfolio will include the following:

- detailed maps describing the exact location of the repository;
- the Safety Analysis Report, the Final Environmental Impact Statement;
- the RCRA Permit and the Compliance Certification Application;
- environmental and ecological background data collected during the pre-operational, the disposal and decommissioning phases;
- records of the waste container contents and locations within the repository;
- drawings defining the configuration of the repository and shafts;
- reports describing how the waste was emplaced and how the repository was decommissioned, closed and sealed;
- design information for the passive institutional controls.

The information will also be distributed to appropriate organisations for long-term safekeeping:

- federal and state government agencies;
- federal, state, tribal, and local archives and libraries;
- local and state record repositories;
- national archives and libraries of nations that possess nuclear weapons and nuclear energy or produce natural gas and oil resources;
- professional and technical societies.

The DOE intends to submit WIPP records to over 100 archives nationally and internationally. The initial submittal of these records will occur after closure and decommissioning of WIPP. It is important to note that final distribution of data to archives nationally and internationally is not foreseen to take place until 2093. This underlines the necessity for having flexible, well-structured databases which can be adapted to changes in technology.

How long are these PIC measures expected to be effective? Based on the characteristics of the markers, the WIPP designers believe that these components have the capability of lasting in excess of several thousand years. The multiple copies of the records in the records centres and archives, the selection of highly durable materials (that is, archival paper and carbon-black ink), and the fact that the records will have value in the economic and health areas, suggest that at least some copies of the records have a high probability of surviving for many hundreds to thousands of years. The WIPP Markers Panel concluded that the messages proposed have a high probability of being understood by all potential levels of technology for at least 2000–5000 years.

11.6 Broad Conclusions

The questions posed at the opening of this chapter were:

- Why and for whom should records be preserved?
- For how long should and can record-keeping systems function?
- How and where can they be reliably preserved?
- What research has been done in this area and what remains to be done?

The responses to the first three questions are effectively common to all repository programmes that have addressed the issues, and are also recorded as a recent IAEA consensus (IAEA, 1999c). Records are meant to provide future generations both with information and with a warning. The purpose of the warning is to prevent or reduce the probability of unintentional intrusion into a waste repository. More detailed information can reduce the consequences of any intrusion, whether unintentional, or deliberately planned in order to retrieve wastes or remediate sites.

Maintaining institutional memory and continuous records for hundreds of years is judged feasible. Countries such as France and Spain, with near surface disposal facilities, include this as part of their basic concept. Archives at local national and international levels should be maintainable for hundreds of years. For very long times, more emphasis is on markers and monuments. These may provide information and warnings for many thousands of years if correctly designed and sited. There has been considerable thought and research devoted to the subject of record keeping by archiving or emplacement of monuments. By far the greatest efforts have been in the US programme, but other countries (e.g. Sweden and France) have also addressed the issues. In addition, countries already disposing of radioactive wastes in near-surface facilities have practical schemes in operation for data collection, archiving and preservation.

Chapter 12

Accounting for Uncertainty

In assessing the potential future behaviour of any technical system, we must always deal with uncertainties. Decisions to implement any technology must always be taken in the light of the residual uncertainties. Often, these can be reduced by extensive prior research and development, by extrapolation from knowledge of similar systems and, ultimately, by observing the performance of earlier examples of the technology in question. This is how society has developed acceptably safe transport systems, power production facilities and drugs, for example. For radioactive waste disposal, however, there are some especially challenging aspects of uncertainty. Whilst there are important uncertainties associated with many aspects of a repository development programme (e.g. in the inventory of wastes that might be sent for disposal and in the timescale of the programme itself), the main concern is always associated with long-term safety. In fact, the long timescales considered in geological disposal are a key feature making treatment of uncertainties more challenging than for other, sometimes more complex, technological undertakings.

Interestingly, the aspect of radioactive waste management which focussed attention on the long timescales and on the corresponding uncertainties in disposal was our very exact knowledge concerning the times taken for radionuclides to decay. The precision of measured half-lives exceeds that in almost all other parameters, but the extremely long half-lives of some radionuclides draw attention to the difficulty of assessing how other parts of the system might behave over such times. In other areas, such as disposing of heavy metals, which stay toxic forever, consideration of far future effects was long neglected. Increasingly, however, society is recognising that some technologies introduced by man may have potentially enormous impacts sometime in the future, that the uncertainties in predicting these impacts can be huge — and that decisions must, nevertheless, be taken in the face of these uncertainties. The examples are growing in number, including pesticides, CFCs and other ozone depleting gases, CO_2 and other greenhouse gases and genetically modified organisms. It may come to be recognised that, by their explicit acknowledgement of the uncertainties of future technology impacts and their search for approaches to address these uncertainties, scientists involved in planning radioactive waste disposal have, in fact, played a pioneering role.

It is today accepted that uncertainty is an unavoidable aspect of planning and regulating deep geological disposal programmes. This fact was recognised early in the development of assessment methodology. Twenty-five years ago, Bartlett et al. (1977) noted that:

> Assessment of geologic isolation safety is unique relative to assessments for other engineered systems because some elements of the analysis are not amenable to uncertainty reduction by additional R&D.

Wisely, the authors went on to warn that "*highly sophisticated models ... could create an unwarranted 'illusion of certainty'*". Nevertheless, they were able to propose specific modelling techniques that could address the task of quantitatively assessing some kinds of uncertainty. Their suggestions were to use fault trees, simulation analysis and stability analysis, all of which were recognised to be still at the conceptual phase.

12.1 Development of a Systematic Approach

Soon thereafter, the terms **uncertainty analysis** and **sensitivity analysis** became common in the technical literature on repositories. The IAEA (IAEA, 1981) included these in a glossary in Safety Series 56, noting that estimation of uncertainty and error bands needed *the application of statistical techniques and definition of input parameters in probabilistic form*. However, it was soon realised that a purely quantitative approach was not possible. In an updated performance assessment methodology report (IAEA, 1985), the following causes of uncertainties that could be addressed quantitatively were identified:

a) inability of models to represent the system completely;
b) approximations used in solving model equations;
c) uncertainties in parameters.

In addition, it was recognised that a problem was presented by "*the inherent irreducible type of uncertainty represented by gaps in our current understanding of the system*" and the opinion was offered that "*there is little that can be done to resolve this type of uncertainty*".

It is worthwhile repeating here that considerations of this sort are, in fact, relevant for many technologies. A more topical and, arguably much more important, example today, concerns the emission of greenhouse gases and their effect on future climate. Great uncertainty in this area results from all three causes discussed above and the completeness of current understanding continues to be hotly debated. A consequence of this is that decision makers do not have consistent guidance from experts or, to be less charitable, they can choose views that support politically motivated ends.

One of the largest sources of uncertainty is predicting the future behaviour of human populations. Societal changes happen much faster than the decades to hundreds of years of importance for climate change mechanisms and very much faster than the tens or hundreds of thousands of years for radioactive waste decay (see Chapter 2). Of critical importance, however, are not the uncertainties in individual parts of the system, but rather the impact of these uncertainties on the overall consequences of any technology. This is one of the key conclusions which have led to geological disposal being favoured for long-lived wastes. Properly chosen deep geological formations are one of the very few environments accessible to man which have such an immensely long stable history that they can reliably isolate wastes from the effects of more transient changes. As pointed out by Thunberg (1999), "*we use the relative predictability of geological time to nullify the uncertainty of other time spans*".

Typically, four types of uncertainty are associated with assessing the future performance of repositories:

- **System uncertainty:** uncertainty as to whether the disposal system (repository, EBS and natural environment) has been sufficiently understood and properly characterised.
- **Scenario uncertainty:** uncertainty as to how appropriate and how comprehensive or complete are the choices of scenarios of future events and processes that will perturb system evolution.
- **Model uncertainty:** uncertainty as to whether the conceptual models used to describe the behaviour of parts of the disposal system represent reality sufficiently well and whether the algorithms of the calculational models correctly represent the conceptual understanding.
- **Parameter uncertainty:** uncertainty over the specific parameter values and parameter ranges to use in the models; these parameter uncertainties may be due to the natural variability of the system or to the inexact nature of our measurement techniques. Of particular importance for repositories are the uncertainties in parameters characterising the geological environment. A large volume of rock must be characterised, the spatial scales of key features determining behaviour are small, and the requirement to leave the natural rock barrier intact precludes extensive destructive testing.

Ten years ago, it was believed by many that most types of uncertainty could be reasonably managed by a combination of parametric sensitivity analyses, probabilistic analyses, model "validation" and the use of alternative conceptual models. With time, however, it has become appreciated that, in assessing uncertainties associated with looking into the far future, other less quantitative approaches are also needed. In 1991, an NEA expert group (NEA, 1991), whilst recognising that uncertainties can never be completely eliminated, expressed the opinion that "*by using both quantitative methods and expert judgement, the amount of uncertainty can be evaluated and* ***a basis for decisions can be provided***".

The key element is the recognition that the goal is *sufficient* confidence in the reliability of the analysis to allow an adequately justified decision to be made. This was made clear in a specific regulatory context by the publication in 1997 of an IAEA report entitled "Regulatory decision making in the presence of uncertainty in the context of the disposal of long-lived radioactive wastes" (IAEA, 1997d). The task of "increasing confidence" in analyses of repository behaviour has continued to grow in importance. In 1999, a further publication entitled "Confidence in the Long-term Safety of Deep Geological Repositories" was published (NEA, 1999a), concluding that "*methods exist to evaluate confidence ... in the inevitable presence of uncertainty*" and giving the specific examples of these methods that are described below. Subsequently, the NEA initiated a dedicated Forum on Stakeholder Confidence (NEA, 2000a).

Although these are all laudable efforts on the part of the waste management community to address the crucial issues of reducing uncertainties and increasing confidence, there is an undoubted element of "preaching to the converted". In the past, outsiders — including even scientists from other disciplines — were often not included in the debate. Researchers pointed out that increasing study of an area could also lead to growth in uncertainty, e.g. because new complicating effects are brought to light. An example is the discovery of plutonium in groundwater at a far larger distance from underground weapons tests than the most commonly accepted scientific theories would have predicted (cited in NRC, 2000a). Furthermore, some controlled experiments aimed at determining how well scientific experts could themselves subjectively estimate uncertainties in knowledge have indicated that the more familiar the expert is with the problem, the greater their estimate of uncertainty is likely to be.

In the light of this complex situation, advice from international agencies is now recognising that regulations need to be discriminating and practicable in what they require of an implementor as assurance of longer-term safety. For example, it is now a principle of geological disposal that *absolute assurance* of safety cannot be achieved: what is sought by regulators is *reasonable assurance of safety* or *reasonable expectation* that the system will perform safely (IAEA, 1997d). The same reservation is obviously valid for any technological endeavour, but in other areas it is not highlighted in the same way. The IAEA recommends that regulations and standards should include a statement to the effect that absolute proof is not to be had, and that the implementors need only provide reasonable assurance of safety, based on the record of information available to regulators and the public. Already in 1993, the regulations in the USA contained explicit reference to this:

> Proof of the future performance of engineered systems and the natural geological setting over time periods of thousands of years is not to be obtained in the ordinary sense of the word.
> ...the standards must accommodate large uncertainties. These include both uncertainties in our current knowledge about disposal techniques and inherent uncertainties about the distant future. (EPA, 1993)

Recent regulations for the proposed Yucca Mountain repository continue with these messages. The USNRC 10 CFR 163 section 101 notes the following:

> Although the performance objective ... is generally stated in unqualified terms, it is not expected that complete assurance that the requirement can be met will be presented. A reasonable assurance ... is the general standard that is required.

The USEPA in its corresponding regulation, 40 CFR 197, section III.C.3, uses a similar term:

> ...we are proposing the concept of "reasonable expectation" to reflect our intent regarding the level of "proof" necessary ...We intend for this term to convey our position and intent that unequivocal proof of compliance is neither necessary nor likely to be obtainable

The back-drop to these considerations is that all parties must be confident that the uncertainties inherent in geological disposal are not so significant as to call a particular course of action into question. The *precautionary principle*, on which (together with the *sustainability concept*) much environmental legislation is now being based (see Chapter 3), implies that, where there is significant uncertainty and a potentially serious risk associated with a new practice, then the practice should not be undertaken until the uncertainty has been addressed. Regulators must thus be confident that uncertainty has been addressed adequately, and/or be prepared to make arguments about the risks of taking alternative measures and about the deployment of society's resources (see Chapter 3).

The remainder of this chapter examines ways in which reasonable assurance of safety might be demanded of the implementor in regulations.

12.2 Providing Reasonable Assurance of Safety

A comprehensive PA will endeavour, as a matter of course, to quantify the impacts of as much of the known types of uncertainty as possible, by means of:

- sensitivity analysis (to parameter variation), which could involve probabilistic analysis;
- the analysis of a range of scenarios of alternative future states of the system;
- the application of alternative conceptual models of features or processes.

To appreciate the importance of the latter two points, it is essential to emphasise again that a single, accurate prediction of future system behaviour is not needed to assure safety. If we look at alternatives, all of which indicate that adequate safety will be provided, then the only question is whether these properly scope the range of possible futures. This brings us again to the question of completeness. Are there effects or consequences which have not been thought of? The uncertainties raised by this question cannot be quantified.

There are, however, also qualitative approaches to reducing uncertainties which complement the more quantitative approaches listed above. Taken together, the different approaches can reduce uncertainties to levels that justify decision making. The various approaches to reducing uncertainty — or conversely to increasing confidence — in the safety of deep geological disposal have been discussed in international fora, such as the IAEA and the NEA (IAEA, 1997d; NEA, 1999a; NEA, 2002a). They are outlined below.

Apply good science and continue well-chosen R&D activities throughout the repository development programme Sound science must underlie the development of safe repositories. Over the past 25 years, extensive scientific work has been done in the wide range of disciplines needed for disposal. Although many workers in the field are of the opinion that most of the basic scientific work has been accomplished, all agree that there are areas which will benefit from more research and all recognise that a wealth of good applied science will be needed for implementing repositories and assessing their performance. An example of an area still requiring research is two-phase flow and transport of radionuclides; an example of a mature area with little need for more work is heat transport through rocks. It should be noted that continuing R&D does not imply that an implementor knows so little about the system that progress towards repository implementation is impossible. Rather, it acknowledges that, over the decades of development, science and engineering will develop, so that approaches can be improved or optimised during this period.

It has been argued, moreover, that waste management scientists could and should draw more on the general scientific knowledge and experience in other fields, e.g. the complex flow and transport processes of interest have long been studied in the oil and gas industries. The particular types of R&D activities most commonly viewed as being of continued importance are those involving large-scale field tests or underground laboratory experiments. This type of work is valuable for reducing uncertainties by bridging the gap between small-scale laboratory experiments and repository scale effects that are not observable because of the long timescales involved.

Use robust designs and analyses (see Box 11) Repository designs should, as far as practical, incorporate some level of obvious conservativeness or even pessimism in order to accommodate uncertainties. This means using designs and materials that are known to be resilient to a broader range of conditions than reasonably expected: effectively, a margin of safety. One illustration of this approach is taking careful measures to seal galleries or shafts which are, in any case, located relative to the existing groundwater gradients in such a way that they would not represent preferred flow paths. However, care must be taken not to expend resources that are unjustified, and the "tolerability of risk" level of 10^{-6} per year (see Chapter 5) is often used as a cut-off, below which it is not considered worth spending more money to reduce risk. Analyses can incorporate similar robustness by ensuring that the

Box 11: The Meaning of Robustness

The word "robust" has been applied to two different aspects of geological disposal (McCombie et al., 1991). In both cases, the fundamental concept is that the outcome of an activity should not be sensitive to uncertainties in the data and assumptions that are used as input. The first activity is the design of the repository system itself; the second is the analysis of the potential future behaviour of that system.

A robust **repository system** is based on:

- simple geology, physics, chemistry, design, which enhance understanding and transparency;
- large safety factors in the individual components, such as large corrosion allowances in choosing waste container wall thicknesses;
- some degree of redundancy in the safety barrier system.

A robust **performance assessment** of the resulting system ensures that:

- the models employed are well validated;
- the models and data are realistic or conservative;
- all potentially negative processes are analysed;
- the calculated results are insensitive to reasonable parameter or model changes.

full range of possibilities is explored for negative factors about which there is uncertainty, and by not taking any credit for positive factors where there is uncertainty. An example is the common neglect in safety assessments of the ability of corrosion products from deteriorated waste containers to adsorb radionuclides.

Aim for simplicity This approach is closely associated with the previous discussion on robustness. A simple safety concept combined with simple PA models can provide considerable insight into how a system functions, and thus enhance confidence. Although complexity usually increases as a system becomes better characterised, and complex models will always be necessary in a repository development programme, a goal should be to achieve some unified simplification that incorporates both considerable knowledge and insight. This is exemplified by the copper container concept for spent fuel that is utilised in Scandinavia. Extensive and complex R&D supports the understanding of how the copper canister — bentonite buffer EBS works, but the concept is simple: total containment until the levels of radioactivity of the spent fuel are the same as the original ore. A further example is the proposal to seek simple, high-isolation sites for geological disposal by performing a worldwide search for environments where the geological and hydrogeological conditions are particularly simple, which would reduce uncertainties concerning the safety of disposal (Black and Chapman, 2001; Miller et al., 1999).

Use a structured approach including iterative assessments The total approach to designing and siting a repository so as to assure that safety is provided — and can be adequately demonstrated — should be laid out clearly and understandably. The key points to be documented and communicated are the elements defining the safety concept and the features and processes involved in making the safety case. The choice of scenarios analysed must be justified and the analysis methodology clearly explained. Several iterations of performance and safety assessment should be an essential component of a repository development programme. These highlight areas of uncertainty and their impacts in the light of growing knowledge about the concept and, eventually, about a site. Whilst they do not in themselves reduce uncertainties, structured assessments provide a framework for analysing and managing them.

Use multiple lines of reasoning, a range of models and natural analogues The results of any particular quantitative model of repository behaviour will not, on their own, give all stakeholders the required level of confidence in system safety. It is necessary to support the interpretations and forecasts within a PA or safety assessment with a combination of alternative predictive models, detailed system models and broad-brush "insight" models based on different principles, and with independent evidence such as that derived from studies of natural analogues (see Box 5). Typical areas where convincing information can be won from studying natural systems include the estimation of corrosion rates of metals and glasses in the ground and the demonstration of low solubility for key radionuclides under repository conditions. It is reassuring to all stakeholders if similar conclusions on the role and impact of a process can be reached using independent sources of evidence.

Document the elicitation of expert judgement It is increasingly recognised that many decisions on the use of information and on the content and scope of PAs must be based on expert judgement. Because this is itself a significant source of uncertainty, the basis of decisions must be well documented and traceable. Techniques have been developed for eliciting expert judgements on parameters or processes in a formalised way that minimises biases caused, for example, by participants being influenced by the group dynamics of a common meeting. The ultimate application of human judgement in this area will, of course, be in the licensing process itself. Reaching a regulatory decision cannot be achieved by application of a simple formula: it will always be a matter of judgement.

Perform quality assured analyses and have these peer reviewed Peer review helps to identify uncertainties. Alternative opinions of experts, who have not been involved in a programme, test both the concepts and the analyses carried out. In order to work well, the programme that is being peer-reviewed must have had good quality assurance, so that all decisions and data are traceable for the reviewers. This allows real uncertainties to be identified directly, rather than being hidden in questions about where information came from or what is assumption and what is fact.

Encourage international cooperation and evaluation This is closely connected with the previous point. In some cases, expertise is spread so thinly throughout the world (e.g. on thermodynamic databases) that international teams are necessary to attack a problem. Some of the experimental approaches useful for reducing uncertainties (e.g. large-scale tests underground) require such extensive resources that single national programmes can hardly afford to work alone. Lastly, in areas where parallel approaches are feasible (e.g. model development, sorption measurements), then the independent results can be compared within a framework of international cooperation, thus increasing confidence in the work. However, as with all approaches dependent upon insiders reaching a consensus on scientific issues, one must guard against this leading to a degree of over-confidence not shared by wider groupings of stakeholders.

12.3 Possible Approach to Uncertainty in Developing Regulations

Regulatory decision-making for long-term repository safety will have to be carried out using a wide range of information, some of which will inevitably be clouded by significant uncertainty. The IAEA suggests that regulatory standards need to begin with a statement that acknowledges this. Such a statement would also indicate that it is expected that some uncertainties will increase when considering times far into the future. This implies that the implementor and regulator would both be expected to adopt a different approach to evaluating performance in the long term and to reaching decisions on acceptability. For example, a common approach is to use stylised scenarios and analyses for the longer term. It would be the judgement of the regulator as to whether such stylisations were acceptable (NEA, 1999b). The options for doing this are discussed in more detail in the earlier chapters on timescales and performance measures, and are not repeated here.

Nevertheless, the precautionary principle (see Chapter 3) is explicit in requiring uncertainties to be addressed as comprehensively as possible in reaching a decision whether to proceed with disposal: simply acknowledging that uncertainty exists is not adequate. An appropriate regulatory response would be to require the implementor to carry out a comprehensive programme specifically to identify and, so far as possible, to quantify uncertainties and their impacts on performance. It would be useful to the implementor if the regulations were to stipulate that a range of information would be required in order to reach a licensing decision. This could mean stipulating an iterative programme of safety assessments at key points in the implementor's programme, aimed (among other things) at quantifying the impacts of uncertainties, in parallel to the developing design and siting studies. Such assessments might be required to include:

- a structured programme of scenario definition and analysis (addressing scenario uncertainty);
- identification and evaluation of alternative conceptual models of system properties and of processes (addressing both system and model uncertainty);
- sensitivity analysis of parameter ranges and combinations, utilising probabilistic methods as necessary (addressing parameter uncertainty, as well as parameter variability);
- application of diverse arguments and multiple lines of reasoning to support key findings or assumptions;
- a fully traceable documentation system that allows quality control on all data and decisions;
- the findings of peer reviews of critical stages of the implementor's programme: in some countries the regulators organise such reviews for themselves (through IAEA, NEA or independently), funded by the implementors.

In conclusion, three points concerning uncertainties in waste disposal are re-emphasised:

1. Quantitatively and qualitatively assessing the uncertainties in potential future behaviour of a repository is recognised as being of key importance. In its 2001 annual report to the US Congress (NWTRB, 2001), the Nuclear Waste Technical Review Board identified as a priority area the meaningful quantification of conservatisms and uncertainties and encouraged the development of multiple lines of reasoning to support the safety case.
2. Uncertainties are unavoidable in all technological enterprises. Accordingly, decisions in all such enterprises must always be taken in the light of residual uncertainties. The NWTRB summed this up well for the case of a deep repository:

   > The Board recognises that any projection of long-term performance ... is inherently uncertain; eliminating all uncertainties will never be possible (although they may be reduced) ... policy makers can make a decision on whether to recommend the site at any time, depending in part on how much uncertainty they find acceptable.

3. Finally, the regulator must have a considered approach to weighing and assessing the results of the safety-based component of a license application. This means being prepared to say something about the weight attached to uncertainties, and the way in which decisions will be reached taking account of the diverse requirements of short-term safety, long-term safety, deployment of appropriate resources and equity with other environmental regulatory decisions that must also involve uncertainty.

Chapter 13

Chemotoxicity and Radiotoxicity: a Common Framework?

As first discussed in Chapter 1, we believe strongly that principles and standards for regulating radiation safety should be looked at in a broader framework of public health protection measures. Harmonisation of protection across all hazards facing the public should be a societal aim; a good starting point is harmonisation of regulations governing the potential exposure of the public to toxic materials. Harmonisation does not mean identical treatment of all hazards; it aims rather at the development of a common policy framework for risk management.

Today, there is a move towards risk informed regulations. This implies a structured approach to *risk assessment* and to *risk management*. The former is a four-step process involving hazard identification, dose–response assessment, exposure assessment and risk characterisation. These are largely scientific tasks that are carried out by experts in the appropriate field. The risk management part, which includes the area of promulgating regulations, is a policy matter that can only be done using the same administrative and political processes as other policy issues.

This chapter describes and compares how the scientific and regulatory aspects are combined when regulating radiotoxic materials and when dealing with chemotoxic substances. In both cases, the focus in the following discussion is on chronic health effects due to exposures to small doses over long times. Much research has been devoted to learning the carcinogenic and genetic effects of radiation, and much is still to be learned. Chemicals can also cause cancer. For both radiotoxic and chemotoxic materials, measures must be taken to prevent people receiving harmful exposures. The measures in the two cases have, however, been developed largely independently of one another. There is no unified system for comparing the two types of hazards. Standards set for chemicals tend to focus on allowable concentrations in the environment, whereas radiation standards relate directly to the effects on individual people.

It is useful to try to put the two on a common footing. This is the purpose of the present chapter. First, the toxic effects of each are discussed and then the differing approaches to legislating protection of the public are discussed, before finally drawing conclusions on the importance of developing more unified approaches.

13.1 Radiotoxicity

13.1.1 Mechanisms and Effects

The chain of reaction leading to a biological effect of radiation is complex. Ionising radiation can directly damage the DNA in the cells of the human body or it can have an indirect effect by altering the chemistry. The same primary event, e.g. a broken chromosome, can lead to different outcomes: cell death, sterility, metabolic changes or even death. There may, however, be no harmful effect, since the body also has natural repair mechanisms. At high doses (in the Sievert range), acute damage can occur through radiation. Above a certain threshold value, the effects are deterministic and depend on the dose level. At the low doses of interest here, the effects are stochastic, which means that the dose received affects the probability of health effects arising (in this case, mostly induction of tumours). The assumption made today is that this probability decreases linearly with dose but that there is no threshold below which no effects arise. This assumption is based on extrapolations from high doses. The predicted effects are so small that they cannot be directly proved or disproved by epidemiological studies and the true dose–effect relationship is an issue that is hotly debated in scientific circles. More controversial is the debate

on the functioning of repair mechanisms and whether low doses can indeed encourage these mechanisms and thus have a positive effect (called hormesis).

How has the knowledge basis on radiation effects been built up? As for other toxic materials, one source of information is from animal toxicology experiments. In the case of radiation, there is also an extensive database on humans exposed to high doses from nuclear weapons and from medical exposures, and from workers in relatively high radiation environments (nuclear facilities, mines etc.). Cancers induced by radiation cannot be distinguished from spontaneous cancers, so that, in all studies, there is a problem of distinguishing background effects. The organs that are most susceptible to radiation-induced cancers are the bone marrow, breast, lungs, stomach, intestine and thyroid. Most solid tumours have a latency period of around 10 years, although radiation-induced leukaemia can occur after around two years.

Quantification of the dose–response relationship for radiation is needed for risk assessment. One advantage when considering radiotoxicity is that effects do not depend on details of the radiation source, as opposed to chemotoxicity, which must be individually examined for all chemicals. For radiation it is common to use a dose to risk conversion factor giving the annual, or the lifetime risk for a specific exposure. The currently accepted risk value given by the ICRP (ICRP, 1991a) is ~0.07 per Sievert. This implies that exposure at the limit commonly proposed for repositories (0.1 mSv/a) gives an annual risk of about 7×10^{-6}/a or about 7×10^{-4} per lifetime (assuming a rounded figure of 100 years: see also, Section 6.1.4).

13.1.2 Current Status of Understanding on Radiological Effects

Making proper use of available scientific evidence in the regulatory process necessitates having a grasp on the level of residual uncertainties. For radiation protection, a concise recent review (NEA, 2000b) by a Committee of the NEA gives a useful overview of the current scientific status. The key points are extracted here:

- The chief somatic effect of ionising radiation at low doses is the induction of cancer. At high doses, greater than 500 mGy, deterministic effects (such as erythema, cataracts, infertility) are known to occur. There is firm evidence of radiation-induced cancer risk in humans at acute doses in excess of 200 mGy.
- Radiation-induced, solid cancers have a long latency period, generally greater than 10 years. Leukaemia and thyroid cancer in children can appear as soon as a few years after exposure. Various host factors (such as age at exposure, time after exposure, gender, genetic predisposition, etc.) and environmental factors (such as cigarette smoking, infectious agents, etc.) influence cancer risk at exposure levels where radiation effects have been observed.
- Cellular repair mechanisms are known to exist. However, misrepair and residual DNA damage occur. No positive biological effects have been observed in humans exposed to acute doses of ionising radiation.

- Epidemiological studies alone will not provide definitive evidence of the existence or non-existence of carcinogenic effects due to low dose or low dose-rate radiation. The lack of epidemiological evidence for the existence of low dose and low dose-rate radiation induced effects is not proof that such effects do not exist. Epidemiological studies have not detected hereditary effects of radiation in humans with a statistically significant degree of confidence.

The NEA report then goes on to list the points that remain unresolved at the low doses and dose rates of interest in radiation protection:

- The mechanism of carcinogenesis, whether induced by radiation or by other agents, is believed to be a multistep process that is not fully understood. Although damage to DNA is assumed to be a key step in radiation carcinogenesis, it is not known what critical lesions in DNA are responsible for gene or point mutations and chromosomal aberrations leading to cancer.
- The shape of the dose–effect relationship at low doses and dose rates for radiation carcinogenesis in humans is in question.[22] The roles of host factors (such as age at exposure, time after exposure, gender, genetic predisposition, etc.) and environmental factors (such as cigarette smoking, infectious agents, etc.) as determinants of radiation risk are uncertain. The basis of biological effectiveness of different radiations (alpha, beta, gamma, neutron) at inducing late effects in humans at low doses and low dose rates are not yet sufficiently understood.
- The influence of repair processes on human radiogenic risk at low dose and low dose rate is not fully understood; however, biological and chemical repair processes of radiation damage are known to occur in cells. It is unknown whether adaptive response, observed in single cells under certain conditions, influences radiogenic risks in humans. It is unclear whether positive biological health effects of low doses of radiation exist in humans.

It is interesting to compare this scientific status of knowledge on radiation with that on the effects of toxic chemicals. This is done below, before summarising the commonalities and differences in approaches to protection.

13.2 Chemotoxicity

Toxic chemicals can act in many ways like radiotoxic materials. They can have local or whole-body effects, cause acute or chronic harm and affect certain vital organs more strongly: in the case of chemicals, often the liver or kidneys. However, the mechanisms for cancer induction by chemicals are far less understood than are those for radiation.

---

[22] The alternatives are (a) threshold, (b) linear non-threshold, and (c) j-shaped or hormetic. Although radiation protection conservatively uses the first, strong arguments have been made for the hormesis theory (Calabrese and Baldwin, 2003).

Carcinogenic chemicals are divided into two groups:

- *genotoxic* chemicals, which interact directly with DNA in the nucleus of cells;
- *epigenetic* carcinogens, which interact indirectly, causing metabolism changes that can lead to tumours.

The latter have an effect only above a dose threshold, the former are assumed to have no threshold. Most carcinogenic chemicals affecting humans are genotoxic. An important difference between radiotoxic and chemotoxic materials is that for the latter there are usually a multitude of underlying mechanisms, in contrast to the relatively well-understood effects of radiation. For chemicals, metabolism plays an important role, in that it may not be the original chemical, but a further reaction product, which leads to cancer.

In principle, data on the toxic effects of chemicals can be derived from animal toxicology experiments, or from epidemiological studies on exposed populations (e.g. occupationally exposed workers). A further possibility is the use of physiologically based pharmacokinetic models, based on understanding the actual mechanisms by which a given agent affects the body. This last method has gained in popularity because the large numbers of agents on which data are needed rules out experiments for each and every one (Seitz, 1998). However, in practice, most data arise from experiments with animals, and a major problem is the applicability of results to humans. A common assumption is that the effect of a toxin is related to the dose per unit body surface area. A safety factor, normally around 100, is then applied to derive values (e.g. acceptable daily intakes) for use with humans. The other problem, as in the case of radiation, is the extrapolation to low doses. As described below, the earlier approach to chemicals, which assumed that they became toxic only above critical doses has been replaced in many cases by a no-threshold approach, as used for radiation.

Findings on one chemical carcinogen are not generally applicable to others, unless similar structures and structure–activity relationships can be relied upon. Accordingly, extensive databases on the toxicity of different chemicals are needed. One example of this is the IRIS database (EPA, 2001b) maintained by the US Environmental Protection Agency. This gives reference oral doses and inhalation concentrations for chemicals; below these values, no adverse effects are expected to occur. It also gives carcinogenicity assessments for chemical agents. For low doses, the extrapolation approach proposed is not simply by fitting to higher dose data. Instead, a linear assumption is recommended, unless there is hard evidence otherwise, and for each chemical, a slope of risk against dose is given.

In summary, it is clear that building an adequate scientific understanding of the toxic effects of all potentially carcinogenic chemicals is a more challenging task still than for radiation. The work began later, the diversity of toxins is much greater and the combined effects of simultaneous exposure to different chemical carcinogens has barely been addressed. The problem of quantifying risks at low doses is very similar for chemicals and radiation. In both cases, the current approach is to favour a linear, no-threshold approach. In both cases, this is an assumption, and a definitive proof of this, or any different theory, will remain difficult or impossible because direct epidemiological trials at the right doses would require huge homogeneous study.

Although the scientific challenge will remain important, the regulatory issue may be easier to solve. As described below, if risk management policy allows recognition of negligible levels of risk, then standard setting becomes feasible, even in the light of residual uncertainties on low dose effects.

13.3 Approaches to Regulations

There are many similarities in how society regards radiotoxic and chemotoxic materials. In both cases, the key issues to be considered when formulating regulatory standards are (Locke et al., 1998):

- the common goal of public health protection;
- focussing on carcinogenic potential;
- the problems of extrapolating to low doses;
- the difficulty of integrating evidence from epidemiology and animal toxicology;
- the major societal issues involved;
- the increasing attention to long-term effects.

In spite of these commonalities, the approach to developing standards has been radically different. In radiation, there is an international "top down" approach based on overall dose or risk limits together with a requirement to reduce exposures if feasible, even below the limits (ALARA). The doses from all sources tend to be integrated, with attention focussed on the recipient. For chemicals, the approach is "bottom up". The tendency is to have source-based individual limits for particular chemicals. Although individuals may be exposed to dozens of chemical carcinogens there is normally no attempt to look at cumulative effects. For radiation, on the other hand, the established approach is to divide overall safety limits down into smaller levels (dose constraints) on the assumption that an individual could be exposed to more than one source.

A further difference concerns the setting of levels of concern as a function of background levels. In radiation protection, the dose constraints set (typically for a repository, 0.1–0.3 mSv/a) are much below the world average natural background radiation level of 2.4 mSv/a. For chemicals, there was a tendency to regard natural levels as negligible, or *de minimis*. Even today, doses that some members of the public receive routinely from environmental contaminants such as lead are only around a factor of 10 below the allowable intakes. On the other hand, when artificial chemicals were first introduced into foodstuffs, there was an attempt in the USA (the Delaney Clause of 1958) to rule that the allowable levels of carcinogens should be zero. As analytical methods improved, this was recognised to be unrealistic, and the law was repealed.

When one moves to compare the actual numerical limits commonly set for exposures to radiation and carcinogenic chemicals interesting differences are observed. For radiation, limits are set and are expected to be strictly observed, with legal repercussions for exceeding them. The limits themselves have been set over the years, using arguments based on scientific observations on exposed persons and comparisons with background radiation and with other societal risks (see Box 2). The typical repository relevant constraint of 0.1 mSv/a corresponds to a lifetime risk of about 7×10^{-4}. To put this in perspective, the *annual* (non-radiological) risk to workers in "safe industries" is 10^{-4} (ICRP, 1977).

For chemicals that are intentionally introduced to food stuffs, stricter risk goals are aimed for, in the range of 10^{-4} to 10^{-6} per lifetime. The crucial difference, however, is that these limits are not regarded in the same binding manner. In particular, for the case of existing chemical contaminants, a less stringent level is often accepted if the risk goal is not achievable on the basis of technical feasibilities or of costs. For example, many cities in the USA fail, in practice, to meet the officially promulgated clean air standards.

Examples of carcinogenic chemicals for which the risks have been assessed (Bossart, 1997) are aflatoxin, a naturally occurring genotoxic carcinogen found in maize, almonds and peanuts, and benzene, a widely distributed air contaminant resulting from motor fuel and industrial processes. The following figures give some useful comparisons of lifetime risks.

- typical dose limit for a repository (0.1 mSv/a) — 7×10^{-4}
- typical limit for radon in houses (1000 Bq/m^3) — 1×10^{-1}
- common measured concentration in air (benzol) — 10^{-4}
- internal body radiation (mostly from ^{40}K in the human skeleton) — 10^{-3}
- USA limit for aflatoxin in foodstuffs — 10^{-5} to 10^{-4}

The overall conclusions that can be drawn concerning regulatory risk limits are that the goals for chemicals are generally lower (10^{-6} to 10^{-4} per lifetime) than for radiation (10^{-4} to 10^{-3} per lifetime). The actual risks to which society is exposed in practice from either is less than 10^{-4} per lifetime. Whether such risks are acceptable to, or accepted by, society has been the subject of much debate. Suffice it here to note that they are significantly lower than many other involuntary risks to which society is subjected, and hugely below voluntary risks to which individual members of the public subject themselves.

Illustrative examples of these other risks (based on data from Fritsche, 1992) are shown in the following Table 13.1.

Consideration of the risks to which society is exposed or to which individual members choose to expose themselves is an essential part of risk management. It should make clear that the appeal for zero risk in any undertaking is fruitless and should help focus a rational debate on risk reduction and optimisation. If the experts and the public could agree that there are incremental risk levels below which we need not discuss further (negligible risk), this could defuse the insoluble problem of extrapolation to minute doses. Unfortunately, academic purists amongst the experts, and scaremongers amongst others, have prevented this happening as yet.

Table 13.1. Lifetime (assuming 100 years) risk of death to an individual

Voluntary risks		Involuntary risks	
Hard drugs	1.0	All cancers	0.78
Smoking 10 cigarettes/day	0.5	All falls	0.023
Helicopter pilot	0.38	Accidents at home	0.0085
Hang gliding	0.1	Pedestrians	0.0029
Coal miner	0.038	Drowning	0.0019
Driving a car	0.01	Electricity	0.00037
Playing football	0.004	Food poisoning	0.00012
Office worker	0.0037	Lightning strike	0.00005

For comparison, a radiation dose of 0.1 mSv/a corresponds to a figure of 0.0007 on this scale.

Given the different origins of regulatory work on chemicals and radiation, an interesting question is whether the end result does reflect any harmonisation of the risk management. Overy and Richardson (1995) looked explicitly at the USEPA regulations when examining this point. They conclude that despite a rather varied approach to setting the standards in the regulations, the end result is that for both chemical and radiological carcinogens, the individual lifetime risk is no higher than 10^{-4}. This is below the lifetime risk of 7×10^{-3} corresponding to the overall ICRP recommended dose limit of 1 mSv/a, and the assertion is that it is highly likely that an individual exposed to a number of sources will still be under this limit.

13.4 Concluding Observations

A good overview of the main similarities and differences between chemical and radiological risks is in the work of Tran et al. (2000). Although the discrepancies between approaches are not major, there are significant differences that could be reduced by further work towards harmonisation. In continuing work on standard setting, it is essential continually to take account of our introductory point to this chapter. Setting standards requires the proper integration of both, scientific work and policy decisions.

Hansson (1998) makes the point that the step of translating scientific knowledge into regulatory standards also requires consideration of the residual scientific uncertainties. He points out that errors can be due to false interpretation of observed facts or to incomplete knowledge, and argues that, whereas the former is of more importance to scientists, in health protection it is crucial to allow for the latter. This argument, that unknown but potentially negative effects must be considered, is allied to the precautionary principle mentioned in Chapters 2 and 3. It must be treated with care, however, since societal considerations, in particular economics, place a limit on how much conservatism we can afford to introduce to compensate for possibly incomplete knowledge.

Standards can only be set by using human judgement on the importance of such unquantifiable issues. This is sometimes the judgement of individual policy makers. It can also be based on eliciting the views of the public, for example by the rule-making processes used in the USA. The judgement may also consider other factors, such as whether the observed public fear of radiation justifies stricter regulation. This can also be a valid approach — but, if taken, it should be done openly. The process of integrating science into standard setting is sensitive. Policy makers must realise that science alone can rarely, if ever, provide an irrefutable answer to where limits should be set. They should also realise, however, that the science cannot be ignored or misused.

Chapter 14

Setting New Standards

Looking afresh at what we regard as the main factors underlying principles and standards allows some scope for suggesting a comprehensive structure, which integrates the most useful approaches that we have identified. In this chapter, we advance items that might be considered for inclusion in any new (or revised) standards and regulations for geological disposal, and propose how those items might be structured into an appropriate set of quantitative regulatory guidelines and criteria.

The proposals made should be seen in the widest context of deep geological disposal of all types of long-lived wastes. Apart from some points of detail, they are considered to be applicable to all categories of long-lived waste and all types of terrestrial, deep underground disposal.[23] At the outset, we note that these suggestions concern only the post-closure radiological safety of the repository. Even though a regulator would need to be concerned closely with every step of repository development, it is assumed that other sets of regulations (and, possibly, other regulatory agencies) will be concerned with:

- the operational safety of a nuclear installation (the repository and its associated surface waste handling facilities);
- other industrial safety and non-nuclear environmental matters.

In this context, the national framework also needs to ensure that the operational health impacts of conditioning and packaging the waste to the point where it can be disposed of are equitable with respect to the long-term safety benefits provided by disposal.

The previous chapters have discussed the issues behind major topics that regulations would need to consider and have made recommendations for either

[23]Some comments on groundwater flow environments under the heading of site selection, and on containment times within the EBS would need to be modified in the case of a repository located in salt deposits, with no flowing water and a safety concept which places less emphasis on the EBS.

adopting a particular approach, or for alternatives that could be considered. Here, we bring these recommendations together in an integrated fashion and propose a simple and comprehensive content and structure for consideration for new regulations. In selecting between alternatives, we have tried to choose options that are likely to reflect the directions of future thinking in standards and in radiological protection.

As noted in the introduction to this book, it is important that standards and regulations are developed in consultation with all concerned parties: government agencies, scientific bodies, other non-governmental organisations and the wider public.

14.1 Policy Framework, Process and Regulation

Many issues discussed in the previous chapters concern the making of national policy, as well as impinging on the setting of standards and regulations. Notwithstanding the need to involve all stakeholders, it is appropriate that the specific regulations (designed largely to protect people and the environment from the harmful effects of radiation) be separated from the wider policy framework and from societal decisions on the process whereby that protection should be ensured. These latter aspects must be developed and agreed by the government, the implementor and other organisations. Nevertheless, if the framework and process are well formulated, they will provide a sound basis for any set of regulations.

Consequently, we first discuss issues that ought to be defined in national policy governing a repository development programme to provide both a framework and a context for regulations. These are outlined in Table 14.1.

Some of these points are addressed in the way in which we have structured the suggested set of regulations in the following section.

Table 14.1. Issues in National Policy

Issue	Possible National Policy Approach
Allocation of responsibilities	National policy needs to be clear about the roles of the various organisations that will participate in the repository development programme and should stipulate for which aspects they are each responsible. Responsibilities need to be clearly defined for all activities and roles, including (among many others) selection of the site, evaluation of safety cases and environmental impact assessments, R&D, advising government, deciding planning matters and funding all the parts of the repository development programme.

(*continued*)

Table 14.1. Continued.

Issue	Possible National Policy Approach
Integrated environmental regulation	Radioactive wastes are only a minute component of the potential and actual environmental contaminants and hazards to which people and other biota are exposed and that contribute to environmental health risks. It is suggested that any new set of regulations should be integrated with, or at a minimum compatible with, the approach taken to regulating other types of environmental health risk. Ideally, all practices presenting similar types of hazards (e.g. contamination of air, water or soil with toxic or carcinogenic materials) should be regulated using similar types of standards. Mitigation of risks can then be harmonised to ensure that resources are used in an equitable fashion to protect present and future generations and environment. This means adopting an even-handed ALARA type of approach to risk reduction in each sector of environment and industry, and ensuring that spending to achieve health benefits in any given sector does not become out of proportion with the actual gains; it has been referred to as "risk-informed regulation".
Environmental impact assessment	It is becoming common practice in many countries for any major industrial development to have to submit, and have approved, a comprehensive assessment of all the environmental impacts of the proposal (an environmental impact assessment or statement: EIA or EIS). The radiological safety of the repository is only one component of its environmental impacts and this might be included as one of the elements of an EIA. The regulator and the implementor need to be aware of: • When, in the repository development programme, EIAs might be required (e.g. prior to selection of the final site; prior to sinking of an exploratory shaft; prior to repository construction) • How the requirements and timetable for submission of an EIA might affect the parallel timetable for submission of radiological safety assessments or iterations of PA

(*continued*)

Table 14.1. Continued.

Issue	Possible National Policy Approach
	• How the regulator would be required to factor their response to radiological safety matters into the overall response to the EIA, presumably with input from several agencies.
Nuclear safeguards	Disposal of spent fuel, or other wastes containing significant amounts of fissile materials, will place a requirement on the implementor to maintain the required level of internationally approved nuclear safeguards. These would require rigorous control of the content and location of waste packages at all times, even after emplacement and backfilling. Following closure of the repository, there would be a continued requirement for surveillance of the site to ensure that material is not being removed. Even if there is no immediate plan to dispose of spent fuel in a repository, the possibility that there may eventually be such requirements placed on the implementor should be considered at the outset, along with the potential impacts on repository design, siting, operation and retrievability policy (see below). These requirements would overlap with any standard regulatory requirements for pre- and post-closure monitoring of the repository system. Consequently, possible future national policy directions, and how these may interface with international requirements, should be considered at an early stage of a repository development programme.
Retrievability and reversibility	A decision by policy makers on whether and how waste disposal should allow for reversibility of the operation and retrieval of the wastes at various times over the operational and post-closure life of the repository could affect both design and operation of the facility. Although retrievability is always possible in principle, requirements for direct individual waste package monitoring or for easy retrievability can have radical effects on design and operation of repositories.

(*continued*)

Table 14.1. Continued.

Issue	Possible National Policy Approach
	The regulator needs to develop a response to such a policy in order to ensure safety is maintained at all times. Where different regulators are responsible for the operational and the post-closure periods, then the regulations covering design aspects of retrieval should be designed to support and enhance safety in each period. It is important to demonstrate that the repository will be safe at all stages of its operational life for different scenarios of waste emplacement and partial completion of the repository (as may be envisaged in response to a requirement for reversibility), as well as after closure. Measures to ease retrievability should not have unacceptable consequences on long-term or operational safety.
Programme milestones and associated performance assessments	A repository development programme should develop iteratively in a "staged" manner, and it would be advisable not to be rigid in terms of deadlines at the outset. The implementor will need to begin to specify activities within the programme at an early stage. One aspect of this will be the need to set in place an iterative programme of performance and safety assessments. Review of the safety case and iterations of the quantitative PA would normally be expected to be carried out at intervals in the repository development programme, coinciding with milestones where internal project decisions are required on siting and design aspects of the programme, as well as for licensing steps. They would also be carried out to develop PA methodology in its own right. To ensure the maximum transparency of the repository development programme, and that the implementor and regulator are fully aware of each others concerns and constraints, the regulator should be involved in each of these iterations, even if they are not directly linked to a formal licensing step.

(*continued*)

Table 14.1. Continued.

Issue	Possible National Policy Approach
	Consequently, it should be a matter of policy that the regulator interacts with the implementor in the definition and review of an iterative series of performance and safety assessments. This policy should also state those milestones in the repository development programme where the regulator would formally be required to make a decision on the continuance of the work (e.g. selection of the final site, first waste emplacement, repository closure). It would be useful if the regulations stipulated a minimal or an optimum approach to interaction with the implementor.
	A further aspect of repository development programme milestones is that they may make it feasible for regulations to be developed in a stepwise fashion, rather than being set up comprehensively at the outset: for example, to address the site selection issue first and long-term post-closure safety requirements nearer the time of closure. There are arguments for and against both approaches, but it is considered best for regulations to be as explicit and comprehensive as possible as early as possible so that the implementor is provided with good guidance and a clear target. The goals must be clear, but flexibility should be allowed in the procedures used to achieve them.
	Nevertheless, it must be expected that regulations need to be able to adapt to changing circumstances over the several decades of a repository development programme, without "moving the goalposts" to the detriment of the programme or public safety. Decisions to modify regulations need to be taken after the widest and most open consultation.
Participatory process	Although it would be expected that approval of formal licensing steps is entirely the responsibility of the regulator, it is possible that national policy may also require other groups to participate (e.g. via local public fora or inquiries) in the wider aspects of decision-making associated with important stages in the repository development programme.

(*continued*)

Table 14.1. Continued.

Issue	Possible National Policy Approach
	For example, the regulator may be required to provide policy makers with an opinion on long-term radiological safety aspects of a proposed site at the time of final site selection, but it is unlikely that the decision to proceed at that site would be based solely on whether a safety case meets regulatory targets. However, it may be possible to uncouple this aspect of process from the way in which regulations are constructed. Demonstration of achievement of regulatory requirements ought to be possible regardless of other factors in a decision. A decision may also be required on the extent to which non-governmental bodies or local government bodies can consult the regulator or ask for a regulatory comment or judgement. The regulator himself can also involve other groups directly, for example by allowing periods of public comment on draft regulations or even using the USA process of rulemaking to incorporate wider opinions directly into the setting of criteria. The policy maker must also consider whether to provide direct financial support to affected groups who wish to challenge the findings of the implementor or the regulator. A system of compensation may be developed to recompense the host community for accepting the burden of a major national facility and its associated development work and long-term operations. The regulator may need to be assured that this system has not had any undue impact on the choice between sites where significant differences in long-term safety may otherwise have been a discriminator.

14.2 Suggested Structure and Content of Post-Closure Safety Regulations for a Geological Repository

This section outlines, in Table 14.2, a possible structure and content for standards and regulations, without providing possible forms of word, text or details. The intention is to identify the main components and concepts, together with any associated quantitative measures that may be needed. This can thus be seen as a template or skeleton upon which to build regulations. Each item would need to be fleshed out and explained in more detail for any national circumstance. It would also be useful for the regulator to provide examples of the types of material they wish to have presented to them by the implementor.

Table 14.2. A model structure and content for post-closure safety regulations for a geological repository

Section	Content
Preamble	Statement of scope, purpose, ethical principles underlying the regulations and approach to compliance: Scope and Purpose • The regulations are intended for the protection of people and the environment from unacceptable health detriments as a result of the harmful effects of radiation • Protection will be afforded to all persons who might be exposed to risk from a repository, regardless of national boundaries that exist now or in the future • The radiological protection principles for disposal of all types of radioactive wastes are the same, although the regulations and criteria may differ (e.g. between surface and deep geological disposal) • These specific regulations apply to any deep repository for long-lived radioactive wastes: where more than one repository or more than one waste type is situated at a single site, the regulations shall apply to the whole site as if it were a single facility Ethical Principles • The repository should adhere to the principle of sustainability, whereby it meets the needs of the present without compromising the ability of future generations to meet their own needs • Future generations and the future environment should be afforded the same level of protection as currently considered acceptable • Impacts at all future times should be considered, although the weight given to different performance measures will vary at different times • Wastes should be managed so as not to impose any burden of responsibility on future generations • Measures to ensure the protection of future generations must not result in an undue health burden to current generations • Freedom of choice for future generations must not be unnecessarily restricted, and no threat of potential harm should be passed to future generations without the possibility that harmful effects can be mitigated

(*continued*)

Table 14.2. Continued.

Section	Content
	• Safety should be seen to take precedence over cost, but the ALARA principle should be applied to ensure that undue resources are not applied to reduce health impacts to unreasonably low levels • The radiological protection criteria for the repository operational phase should be the same as for other nuclear facilities (but these criteria are not treated in these regulations) • The public should be allowed to participate in the decision-making processes, for which demonstration of regulatory compliance is only one contributor • Waste producers/owners are responsible for providing the finances needed to ensure safe disposal/polluter pays principle) Compliance Approach • There can be no absolute proof in a mathematical sense that a repository will behave in any particular way over the very long time periods to be considered • Consequently, compliance with these regulations will be determined on the basis of the provision of reasonable assurance that such protection will be afforded by the proposals presented by the implementor: this determination will be based on judgements that consider a range of performance indicators and a range of approaches to describing future behaviour of the repository
Purpose and design of the repository	A statement and outline of what is expected of the repository, and should thus be seen to underpin its design and safety functions: • The repository is intended to isolate the wastes from the environment, people, and their activities, and protect them from the harmful effects of radiation • The repository should be designed to provide passive safety at all times in the future, without the necessity of active intervention • Protection should be provided using a multiple system of barriers which work together to isolate the waste: safety should not rely entirely upon the functioning of a single component of the disposal system

(*continued*)

Table 14.2. Continued.

Section	Content
	• Any measures taken to allow for continued monitoring of the facility or for easier retrieval of the wastes must not compromise the safety or performance of the disposal system • The repository must provide substantially complete containment of all radionuclides within its engineered barrier system for a period of about 1000 years, to allow for decay of the highly radioactive, shorter-lived fission products
Site selection and characterisation	The implementor should adopt a logical and orderly approach to siting which ensures that: • Natural features assuring long-term safety are present at the final selected site. At a minimum, these should include: demonstrable geological stability and a lack of susceptibility to major environmental change over the next few hundred thousand years; absence of identifiable natural resource potential; a low-energy groundwater system with no fast pathways connecting the repository zone to the surface; conditions favourable for the preservation of EBS longevity; host rocks suitable for repository excavation and engineering • Obviously unsuitable areas are removed from the selection process at the outset, using a set of exclusion guidelines that eliminates or reduces the probability of occurrence of certain disruptive scenarios so that they can be omitted from consideration or be treated in an agreed manner: these guidelines to be agreed in advance with the regulator • The site should have no special features which would enhance the probability of human intrusion. The implementor should demonstrate that preference has been given to sites where such likelihood is minimal • The site can be characterised adequately without excessive disturbance, using currently available technology, so as to provide information adequate to make a safety case with confidence • While it is neither possible nor necessary to identify a safest site, the proponent should demonstrate that no obviously better conceptual options for siting have been overlooked

(*continued*)

Table 14.2. Continued.

Section	Content
Safety assessment	The implementor should set in place a methodology for performance assessment and a planned programme of iterative total system safety assessments at the outset of the repository development programme. A minimum programme would be for submission of safety assessments for review and approval at the following major milestones of the repository development programme: • Comparison of conceptual siting options • At the point of selection of a final site • Prior to commencement of construction of the repository (not including any underground exploratory or characterisation phase) • Prior to first waste emplacement • Prior to closure and sealing of the repository The last four of these steps will be part of the formal licensing process. The safety assessments presented by the implementor should: • Provide information of a sufficient quality, transparency and traceability to assist in making reliable decisions at the relevant point in the repository development programme: in this respect, their depth and detail, together with the acceptable levels of uncertainty that they contain, would be expected to reflect the stage of the repository development programme at which they are carried out • Indicate the safety concept or strategy for the repository and provide reasonable assurance that the required level of protection will be provided to people and the environment, using a range of performance indicators • Provide information on the expected evolution and performance of the repository and on an illustrative range of scenarios covering potentially disruptive events • Use multiple lines of reasoning to support arguments and to provide confidence in the results presented • Explicitly address uncertainties using alternative conceptual models, sensitivity analyses and comprehensive documentation of the use of expert judgement

(*continued*)

Table 14.2. Continued.

Section	Content
Performance indicators	Safety assessments presented by the implementor should include quantitative estimates of the following range of performance indicators, which, together, should provide reasonable assurance of long-term safety for the expected evolution of the disposal system: 1. Radiation doses to individuals at different times after closure 2. Time-averaged flux of radionuclides and other toxic species from the repository to the biosphere 3. Toxicity indices for the waste at different times after closure 4. Radionuclide distributions and concentrations within each component of the multibarrier system at different times after closure Calculations for each of these indicators should be presented for a period of time for which there is reasonable predictability in the expected evolution of the repository environment and, in any case, for a period of 1 Ma. For periods beyond 1 Ma, qualitative evidence, discussion and arguments should be used to illustrate the behaviour of the wastes and the repository. For stochastic, disruptive events with non-negligible integrated probabilities over the 1 Ma period, the implementor should also present calculations of individual radiation doses linked to separate evaluations of the likelihood of exposure occurring. The latter may use qualitative arguments where no reliable quantitative information is available. The level of analysis for each type of disruptive event is to be agreed with the regulator. In all cases, disaggregated dose-plus-likelihood information should be presented together with a discussion of the nature of the event, its causes and environmental consequences. These discussions should consider the varying impacts of each type of event in different time frames: e.g. up to 100 years after closure, 1000 years, 100 ka and 1 Ma.

(continued)

Table 14.2. Continued.

Section	Content
	Individual doses should be calculated using a range of reference biospheres and reference exposure groups, based on current lifestyles in typical environments into which radionuclides may be released, accounting for potential future environmental change. Doses should be calculated for an average individual at the higher end of the range of exposures in the group evaluated. The reference biospheres and exposure groups should be agreed with the regulator.
	For a single repository, a target individual dose constraint of 0.3 mSv/a, as a result of the expected evolution of the repository, should be used as the basis for decision-making. During the period between the total containment time of ~1000 years, up to several hundred thousand years, this figure should be regarded as a regulatory limit that may not be exceeded by calculated values from any plausible release scenario. For times beyond this, performance indicators 2 to 4, defined above, will be used as the principal measures when judging compliance. In this context, a single repository shall be taken to include all wastes disposed of at one site, even though more than one engineered facility may be present.
	Human intrusion impacts (also taken to include other impacts of the activities of people) should be evaluated separately from the expected evolution of the repository and natural disruptive events. They will be considered principally at the time of approval of final site selection. At this time, the implementor should illustrate the impacts of intrusion and other potential human impacts using a small group of scenarios whose nature and content is to be agreed with the regulator. The full range of performance indicators should be used. The scenarios should describe the circumstances under which radiation doses might arise, the groups of people who might be exposed and the sequences of action that might lead to exposure. The implementor should comment on the likelihood of such exposures occurring, providing a disaggregated dose-plus-likelihood description. The scenarios should consider both the direct exposures resulting from intrusion events (to the intruder and to the public), and potential further exposures that might result from damage to the repository barriers.

(*continued*)

Table 14.2. Continued.

Section	Content
	Where calculated individual doses for human intrusion exceed 100 mSv/a for exposures that are prolonged after the intrusion event, the regulator may require measures to be taken to reduce the probability of such doses arising. Since these calculations will be assessed principally at the time of site inter-comparisons, such measures are likely to involve selection of a different site, or a different repository location or depth at a site.
Control and monitoring	The safe functioning of the repository should not depend on continued post-closure monitoring or control of the site or its surroundings. The implementor should: • Provide information to the regulator that demonstrates an understanding of baseline (pre-excavation disturbance) conditions and environmental trends at the chosen site, prior to approval being granted for repository construction. This baseline information should be presented as part of an environmental impact and safety assessment. The regulator will require to evaluate such information before the implementor proceeds with exploratory shaft construction • Establish an operational phase geoscientific and environmental monitoring programme of the repository and its environment that will provide data adequate for making decisions on moving from stage to stage towards closure. The data derived should be provided on a routine basis to the regulator, who will require discussions on the results of monitoring, and on any responses that the implementor may propose, in terms of changes to repository development programme stages, or repository design and layout that may affect long-term post-closure performance. This requirement will be additional to any requirements placed on the implementor by the regulator supervising operational safety • Propose at the outset of the programme a post-closure monitoring concept that can provide relevant data on the behaviour of the repository system, using currently available technology (whilst acknowledging that technology will develop before such a concept is implemented). • At closure, provide a safety case and history of monitoring information that will provide adequate support to justify a decision to seal the repository, and the most appropriate manner in which to do this

(*continued*)

Table 14.2. Continued.

Section	Content
	• Ensure that any necessary surveillance for the maintenance of nuclear safeguards does not involve techniques that will impair the long-term performance of the repository • At the time of closure, put in place on the site durable markers showing the nature and location of the repository, and its original access points
Provision of information to the regulator	The implementor should meet formally with the regulator from the start of the repository development programme to keep the regulator informed of progress and issues arising. The regulator may wish to provide (and publish) guidelines and advice to the implementor at times during the repository development programme, but would only provide formal licensing approvals at the following, pre-defined licensing steps: • Selection of the final site • Start of construction of the repository • First emplacement of waste • Closure of the repository The regulator may require the implementor to stage the build-up to full operation by first operating a pilot facility. During the operational phase, monitoring data should be provided to the regulator, according to a programme to be agreed between the two parties. If significant deviations from expected behaviour are observed, the regulator may require a revised performance assessment to be submitted and approved as a pre-requisite for continued operations.
Documentation, review and quality assurance	The implementor should institute a programme of Quality Assurance that meets international standards, at the start of the repository development programme. All implementor and contractor reports relevant to the site or the repository should be provided to the regulator immediately after they have been approved as part of this QA system. The implementor is advised to publish all such material openly at the earliest opportunity. In addition, the regulator will have the right of access to all information and records in the possession of the implementor, at any time.

(*continued*)

Table 14.2. Continued.

Section	Content
	The regulator will periodically institute independent reviews of the progress of the repository development programme or of key reports or submissions at important repository development programme milestones. These will require the implementor to provide material and briefings to the regulator and any review groups appointed by the regulator. The implementor should, at the start of the repository development programme, put in place a system (replicated at different locations) for maintaining records of all aspects of the repository, its location, design, construction, waste inventory and operational history. These records should be kept using a variety of media and at the end of the repository development programme, should be lodged at multiple locations, nationally and internationally.

14.3 Compliance

A final consideration is the way in which compliance with regulatory criteria should be used in decision making. The suggestions above propose only one quantitative regulatory criterion: the individual dose constraint. It is important to recognise that numerical compliance with the dose criterion should, alone, not compel acceptance of a proposal. The overall quality of the safety case is paramount. In addition, other factors, including policy and societal matters, need to be taken into account by the decision maker. Conversely, exceeding the dose criterion should not necessarily oblige rejection of a proposal, as there may be many unquantified conservatisms present in the safety case. For far future times, it might also be that estimates will exceed the target value as uncertainties increase. However, such cases must be carefully analysed and justified, to see whether additional measures could be taken to improve protection, within the overall ALARA scheme.

Chapter 15

Conclusion

The task of formulating broadly acceptable radiological safety standards for disposal of long-lived wastes is made difficult by three important factors:

- Fear of radiation: the long-standing and well-established public fear of radiation adds a major, perhaps dominating, societal aspect to the scientific challenge. This is in part due to the common public perception of low, "non-natural" doses. Despite being exposed to a pervasive natural radiation background that is, on average, 20–30 times higher than typical regulatory dose constraints for a repository, and hundreds or thousands of times higher than the estimated doses that might arise, many people do not accept that properly managed waste disposal will cause no harm. This is in part due to the prevalence of the linear-no-threshold theory, which assumes that harmful effects of radiation persist, even at levels where they will never be observable.
- A geological repository system, although passive and with relatively few components, incorporates both engineered and natural barriers. The large scale and the spatial heterogeneity of the geological barriers, given that they must be in principle characterised by non-destructive testing, implies that there will always be some uncertainties remaining about their precise properties.
- Long timescales: The fact that a small fraction of the radionuclides in a geological repository will remain radioactive for very long times (greater than 100,000 years) leads to what is, perhaps, the greatest scepticism towards deep disposal amongst the public and some scientists. Indeed, the very attempt to make meaningful statements about the behaviour of a repository over such time frames has been labelled by some as scientific arrogance.

There is little hope of changing people's perspectives on the first factor in the immediate future. The wide-spread fear will only be changed by generations of familiarity with nuclear technology and by a (re)building of trust in the scientific community. Understanding of low doses will improve only by prolonged, wider discussion and application of comparative risks in societal decision-making for environmental protection issues, and by a more even-handed approach by opinion-formers and the media. As we have advocated throughout this book, we believe that understanding of this issue would also be assisted considerably by anchoring radiological protection objectives for waste disposal much more obviously and closely to the widely varying natural background radiation to which we are all constantly exposed.

There are three approaches to handling the incompleteness and inexactness of geological data. The first is to seek host geological environments where these problems are minimised, for example very low permeability homogeneous clays in non-dynamic groundwater systems. The second is to be conservative and take little credit for the radionuclide transport retardation function of the geological barrier; the prime purpose of the host rock then becomes to provide a stable, benign environment for the engineered barriers. The third is to take advantage of the continually improving quantitative analysis capabilities of geology, inter alia by refining probabilistic approaches. From being a qualitative science concentrated on describing the past, geology in its widest sense, has developed into a quantitative and a predictive science.

The answer to the problem of long timescales is that there do exist environments where it is a perfectly sound scientific goal to understand behaviour over hundreds of thousands of years. The deep geological structures of Earth's crust provide such environments — indeed this is the scientific fact which led to development of the concept for deep repositories. In the context of this book, it must also be appreciated that:

- the long timescales involved rule out direct observational proof of repository safety: quantitative modelling is needed;
- the nature of the geological environment rules out highly precise models;
- standards must be framed to allow for the inherent uncertainties that result; they must be based on conservative approaches and reasonable assurance that compliance requirements are fulfilled.

In fact, the central issues in the area of standard setting for disposal revolve directly around the question of compliance requirements. The fact that a 100% proof of safety can never be achieved is common to other technological enterprises. Scientists have developed probabilistic approaches to analyses and corresponding criteria for compliance. Beyond these hard, quantitative approaches, however, the scientists — and even more so, the public — seek reassurance through various qualitative approaches. These are all intended to raise the level of confidence in the safety assessments for a repository.

A common key element in these measures is use of conservative procedures. Design, engineering and construction should all be carried out conservatively. The safety analyses incorporate much conservatism. Specifically, in the context of this book, the principles applied and the standards set should also reflect conservatism. This is indeed the case. Radiation protection limits have reduced over the years by large factors. Radiation protection principles and application of repository safety assessments are both strongly based on conservative approaches which protect future generations as well as current populations.

There must, however, be limits to the degree of conservatism applied. Intragenerational equity considerations imply that a balance must be drawn with other hazards facing current generations. This is the argument that encourages efforts towards harmonisation of regulation. Intergenerational equity considerations limit the burdens which might be put onto current or future generations at the expense of the other. One aspect of "future safety" in a world with a burgeoning population is to use science and technology wisely, so that living and health standards improve for all. This can only be assured if technology continues to improve our prospects in a sustainable fashion. However, some technologies bring with them examples of the very environmental problems with which we are wrestling. Whilst we should not burden future generations, we cannot endow them with a safer future without using technologies today that produce wastes. There is a fine balance between future burden and present necessity. Unrealistic, zero-impact targets, whether for radioactive wastes or other environmental hazards, will not help either our descendants or ourselves.

The principles we adopt today, and the standards we set, directly determine the balance between present and future gains and burdens from any technology. In the nuclear area, there is a very topical reason for being interested in the impact of regulations on developing technologies. Nuclear technologies can be developed in good conscience only if safe methods for taking care of the wastes are available and are endorsed by society. Confidence in safety and sufficiently wide endorsement will be possible only if a clear regulatory framework is established and implemented.

There is another reason why establishing widely agreed integrated standards and approaches to environmental impacts is important. As we write these conclusions, there is a growing acceptance in the scientific community that global warming is real. At the same time, energy supply provision, with rapid deregulation in many countries, has been driven by short-term commercial motives. As a result, there have been localised, but high profile, energy crises in some parts of the world. Five years ago, it would have appeared hopelessly naive to predict any growth in nuclear power in most western countries. Today, nuclear generation, which produces no greenhouse gases, appears to be a much more realistic factor in equations for future global energy supply. This is a bitter cup to swallow for many long-time opponents of the nuclear industry, and there is considerable resistance even to acknowledging this fact, let alone to taking advantage of this source of power. Global warming is an issue where we have the chance to take sensible measures in many of the developed countries, yet there is a disturbing lack of rationality in evidence in many countries.

In this context, we speculate that there is unlikely to be significant new "nuclear build" unless the waste disposal issue can be seen clearly to have been resolved. Evidence of this can be seen in Finland, where the present rapid progress towards development of a deep repository for spent fuel disposal has facilitated the decision in May 2002 to construct additional nuclear generating capacity. An integral part of achieving consensus on nuclear matters is to agree to the principles, standards and compliance requirements associated with waste disposal.

The time for a review of principles and standards is opportune because some countries (such as Finland) are nearing implementation of deep repositories, others (like Canada and the UK) are at crossroads, considering all options, while others (e.g. Czech Republic, South Africa, Slovakia) are just starting out.

In conclusion, we would like to add some general observations for readers to take away from this book.

- A solid, ethically based framework is necessary for formulating standards for radioactive waste disposal. This has been established.
- In setting standards, harmonisation across different activities presenting a potential hazard is an important goal. Only thus can we hope for an equitable use of society's resources.
- We should strive for a responsible use of science in these efforts. Policy should derive from sound science, and scientists — and the result of their efforts — should not be misused in an attempt to justify retrospectively non-technical decisions.
- Nevertheless, we should recognise that science alone does not provide all the answers. Setting sensible standards is an interdisciplinary exercise that must incorporate value judgements about peoples' expectations and the way they want to use their resources. It necessitates direct involvement of wide societal elements if a widely accepted set of principles, standards and regulations is to result.

Our hope is that we have selected in this book a range of waste disposal issues that are topical and important, and that we have made constructive, state-of-the-art suggestions on how to approach these. If we have succeeded in this, then the book may help

- established professionals in the waste disposal field to review the basis, structure and effectiveness of the regulations governing their work and, if necessary, to amend these;
- newcomers to the field to obtain a comprehensive insight of the key issues that they will encounter in this challenging, interdisciplinary and societally very important task;
- the interested public to appreciate and understand the intensive, multi-year efforts that have gone into making radioactive waste disposal a technical area whose principles and scientific methodology may point the way ahead for other modern technologies.

References

Adams, M.R., & Kaplan, M.F. (1986). *Marker Development for Hanford Waste Site Disposal.* Waste Management '86: waste isolation in the U.S. technical programs and public education. Proceedings of the symposium on Waste Management at Tucson, March 2–6, 1: General interest, pp. 425–432.

AEC (1997). *Guidelines on Research and Development Relating to Geological Disposal of High Level Radioactive Waste in Japan*, Advisory Committee on Nuclear Fuel Cycle Backend Policy, Atomic Energy Commission of Japan.

AECB (1987). *Regulatory Objectives, Requirements and Guidelines for the Disposal of Radioactive Wastes.* AECB Document R-104, Atomic Energy Control Board, Canada.

AECL (1994). *Environmental Impact Statement on the Concept for Disposal of Canada's Nuclear Fuel Waste.* Atomic Energy of Canada Ltd, AECL-10711, COG-93-1.

Ahearne, J.F. (1997). Radioactive waste: the size of the problem. *Physics Today, 50*(6), 24–29.

Ahearne, J.F. (2000). Intergenerational issues regarding nuclear power, nuclear waste and nuclear weapons. *Risk Analysis, 20*(6).

AkEnd (2000). *Zwischenbericht, Stand Juni 2000.* Arbeitskreis Auswahlverfahren Endlager-standorte. www.akend.de

AkEnd (2001). *2. Zwischenbericht; Stand der Diskussion.* www.akend.de

Bartlett, J.W., Burkholder, H.C., & Winegardner, W.K. (1977). *Safety Assessment of Geologic Repositories for Nuclear Waste.* BNWL-SA-6068, 1977.

Benford, G., Kiekwood, C.W., Otway, H., & Pasqualetti, M.J. (1991). *Ten thousand years of solitude? On inadvertent intrusion into the Waste Isolation Pilot Project Repository.* LA-12048-MS. Los Alamos National Laboratory.

Berger, A., Li, X.S., & Loutre, M.F. (1998). Modelling northern hemisphere ice volume over the last 3 Ma. *Quat. Sci. Rev., 18*, 1–11.

Berry, W.E. (1983). *Durability of Marker Materials for Nuclear Waste Isolation Site.* ONWI-474.

BFE (1998). *Schlussbericht Energie-Dialog Entsorgung.* Schlussbericht des Vorsitzenden z.H. des Eidg. Departements für Umwelt, Verkehr, Energie und Kommunikation.

BIOMASS (1999a). *Safety indicators complementary to dose and risk, for the assessment of radioactive waste disposal.* BIOMASS Project Working Material, Draft Report, Limited Distribution. IAEA, Vienna.

BIOMASS (1999b). *Guidance on the definition of critical and other hypothetical exposed groups for solid radioactive waste disposal.* IAEA BIOMASS Working Document BIOMASS/T1/WD03.

BIOMOVS (1996). *Development of a reference biospheres methodology for radioactive waste disposal.* Working Group of the BIOMOVS 2 Study. Technical Report 6. Swedish Radiation Protection Institute, Stockholm.

Black, J.H., & Chapman, N.A. (2001). *Siting a high-isolation radioactive waste repository: technical approach to identification of potentially suitable regions worldwide.* Pangea Technical Report PTR-01-01, Pangea, Baden, Switzerland.

BNWL (1974). *High-Level Radioactive Waste Management Alternatives.* 4 Vols. BNWL-1900, Richland, Washington, Pacific Northwest Laboratories, May.

Boursier, F.O. (2000). *Record and archival system for the monitoring period of a surface disposal facility.* Proceedings of Waste Management Symposium 2000, Tucson, Arisona.

Bossart, E. (1997). *Radioactive und chemische Stoffe: Langzeittoxizität, Beurteilungsstrategien und vergleichende Risikobeurteilung.* Unpublished Draft Report, Nagra, Baden, Switzerland.

Brenn, B.J., & McCall, A.M. (1997). *The influence of retrievability on repository design concepts in the UK.* In: Nagra (ed.), Proceedings of the Fourth international workshop on Design and Construction of Final Repositories: influence of retrievability on design, construction and operation of final repositories. Seehotel Kastanienbaum, Lucerne, October 8-8, 1997.

Brenneke, P. (2000). *Recent developments in the German approach to radioactive waste disposal.* In: IBC Global Conferences Ltd (ed.), Radioactive waste management and decommissioning, 3–7 July 2000, Christ's College, Cambridge, 16th Residential Summer School. Course notes. IBC Technical Services, London.

Brocoum, S.J., Van Luik, A.E., Gil, A.V., & Lugo, M.A. (1996). *U.S. Department of Energy Perspective on High-Level Waste Standards for Yucca Mountain.* Spectrum'96, Seattle, Washington, American Nuclear Society. La Grange, Illinois.

Budnitz, R.J., Apostolakis, G., Boore, D.M., Cluff, L.S., Coppersmith, K.J., Cornell, C.A., & Morris, P.A. (1997). *Recommendations for probabilistic seismic hazard analysis: guidance on uncertainty and the use of experts.* U.S. Nuclear Regulatory Commission, NUREG/CR-6372, 2 vols.

Bunn, M., Holdren, J.P., & Wier, A. (2002). *Securing Nuclear Weapons and Materials: Seven Steps for Immediate Action. Project on Managing the Atom.* Belfer Center for Science and International Affairs, John F. Kennedy School of Government, Harvard University.

Busby, C. (ed.) (2003). *Recommendations of the ECRR: The Health Effects of Ionising Radiation Exposure at Low Doses and Low Dose Rates for Radiation Protection Purposes: Regulators' Edition.* Published on Behalf of the European Committee on Radiation Risk Comité Européen sur le Risque de l'Irradiation, Brussels by Green Audit. ISBN: 1 897761 24 4.

Calabrese, E., & Baldwin, L. (2003). Toxicology rethinks its central belief. *Nature, 421*, 13th February.

Cave, L. (2001). A deontological solution to the waste problem. *Nuclear Engineering International*, November.

CEC (1988). *The regulatory framework for storage and disposal of radioactive waste in the member states of the European Community.* Commission of the European Communities. Radioactive Waste Management Series, EUR 11292 EN, Burholt, G.D. & Martin, A., Associated Nuclear Services, Epsom, UK, Publ. by Graham & Trotman.

CEC (1992). *Radioactive waste disposal: recommended criteria for siting a repository. Euradwaste.* Series 6, CEC Report EUR 14598 EN.

CEC (2002). Draft proposal for a *COUNCIL DIRECTIVE (Euratom) on the Management of Spent Nuclear Fuel and Radioactive Waste*. CEC, Brussels.

Chapman, N.A., & McKinley, I.G. (1987). *The Geological Disposal of Nuclear Waste*. J. Wiley & Sons.

Chapman, N.A. (2002). Long Time Scales, Low Risks: Rational Containment Objectives that Account for Ethics, Resources, Feasibility and Public Expectations — some thoughts to provoke discussions. In: *The Handling of Timescales in Assessing Post-closure Safety of Deep Geological Repositories*. OECD Nuclear Energy Agency, Paris, pp. 145–154.

Clarke, R. (1999). Control of low-level radiation exposure: time for a change? *J. Radiol. Protection*, *19/2*, 107–115.

Clarke, R. (2001). *The Role of Ethics and Principles*. In: Better Integration of Radiation Protection in Modern Society. The Second Villigen Workshop. Paper 26. OECD/NEA. Unpublished proceedings.

CNE (1998). *Réflexions sur la réversibilité des stockages* [with an executive summary in English: *Thoughts on retrievability*]. Commission Nationale d'Evaluation relative aux recherches sur la gestion des déchets radioactifs, CNE, Paris.

Cochrane, T., & Paine C. (1998). *Proposal for Augmenting Funding for the Disposition of Russian Excess Plutonium*. Natural Resources Defence Council, Washington D.C., USA.

Cramer, J.J., & Smellie, J.A.T. (1994). *Final report of the AECL/SKB Cigar Lake*, Analog Study. SKB Technical Report TR 94-04, SKB, Stockholm.

Crowe, B.M., Johnson, M.E., & Beckman, R.J. (1982). Calculation of the probability of volcanic disruption of a high-level radioactive waste repository within southern Nevada. USA. *Rad. Waste Manag. & Nuc. Fuel Cycle*, *3*, 167–190.

DEFRA (2001). *Managing Radioactive Waste Safety*. Proposals for developing a policy for managing solid radioactive waste in the UK. Department for Environment, Food and Rural Affairs, London.

DETR (1999). *Report by the United Kingdom on intentions for action at the national level to implement the OSPAR strategy with regard to radioactive substances*. UK Department of the Environment, London.

Dhérent, C. (2000). *The policy of the Archives de France for archiving electronic documents*. In: Proceedings of the DLM-Forum on electronic records. European citizens and electronic information: the memory of the Information Society, Brussels, 18–19 October 1999, Luxembourg, pp. 172–179.

DLM-Forum (1997). *Guidelines on best practices for using electronic information: How to deal with machine-readable data and electronic documents*. Office for Official Publications of the European Communities, Luxembourg. Available at http://europa.eu.int/ISPO/dlm/documents/gdlines.pdf

Dodd, D.H., Heijdra, J.J., & Prij, J. (1997). *A repository design for the retrievable disposal of radioactive waste in rock salt*. Nagra (ed.), Proceedings of the Fourth international workshop on Design and Construction of Final Repositories: influence of retrievability on design, construction and operation of final repositories. Seehotel Kastanienbaum, Lucerne, October 8–8, 1997, p. 2.

DOE (1996). 40 CFR 191 *Compliance Certification Application for the Waste Isolation Pilot Plant*. DOE/CAO-1996-2184. 21 volumes. Carlsbad, New Mexico. U.S. Department of Energy, Carlsbad Area Office, TIC: 240511.

DOE (1998). *Viability Assessment of a Repository at Yucca Mountain*. U.S. Department of Energy, DOE/RW-0508, Office of Civilian Radioactive Waste Management, Washington, D.C.

DOE (1999). Draft, *Environmental Impact Statement for a Geologic Repository for the Disposal of Spent Nuclear Fuel and High-Level Radioactive Waste at Yucca Mountain.*
DOE (2000). *Disruptive Events Process Model Report.* TDR-NBS-MD-000002REV00 ICN01, at www.ymp.gov/documents
DOE (2001a). *Yucca Mountain Science & Engineering Report.* Report No. DOE/RW-0539. (http://www.ymp.gov/documents/ser_a/index.htm)
DOE (2001b). *Yucca Mountain Preliminary Site Suitability Evaluation.* Report describes the preliminary results of the U.S. Department of Energy's evaluation of whether Yucca Mountain is a suitable site for a nuclear waste repository.
Duncan, I.J. (2001). *Radioactive Waste: Risk, Reward, Space and Time Dynamics.* Unpublished D.Phil. thesis, University of Oxford.
EKRA (2000). *Disposal Concepts for Radioactive Wastes.* Final Report, Expertengruppe Entsorgungskonzepte für radioaktive Abfälle. Federal Ministry for Environment, Transport, Energy and Communicatons (UVEK), Bern, Switzerland.
Electric Power Research Institute (1993). *Technical Basis for EPA HLW Disposal Criteria.* Proceeding EPRI Workshop 1, Prepared by Rogers and Associates Engineering Co., EPRI-TRI-10034.
Electric Power Research Institute (2001). *Performance Confirmation for the Candidate Yucca Mountain High Level Nuclear Waste Repository*, EPRI Report 1003032.
Eng, T., Norberg, E., Torbacke, J. et al. (1996). *Information, conservation and retrieval.* SKB Technical Report 96-18. Swedish Nuclear Fuel and Waste Management, Stockholm,
English M. (2000). *Who are the Stakeholders in Environmental Risk Decisions.* In: NEA (2000a) op. cit.
Environment Agency (1996). Radioactive Substances Act (1993). *Disposal Facilities on Land for Low and Intermediate Level Radioactive Wastes:* Guidance on Requirements for Authorisation. UK Environment Agency, London.
EnPA (1982). *Energy Policy Act of 1982:* Section 801: Nuclear Waste Disposal.
EPA (1977). *Environmental radiation protection standards for nuclear power stations* 40 CFR 190, Environmental Protection Agency, Washington D.C.
EPA (1985). *Background Information Document Final Rule for High-Level and Transuranic Radioactive Wastes.* EPA 520/1-85-023. U.S. Environmental Protection Agency EPA, Washington D.C.
EPA (1993). 40 CFR 191, *Environmental Radiation Protection Standards for the Management and Disposal of Spent Nuclear Fuel, High-Level and Transuranic Radioactive Wastes.* Final Rule, Federal Register 50/182, 38066–38089. U.S. Environmental Protection Agency EPA, Washington D.C.
EPA (1998). 40 CFR 194: *Criteria for the Certification and Recertification of the Waste Isolation Pilot Plant's Compliance with the 40 CFR 191 Disposal Regulations:* Certification Decision Final Rule, Fed. Register *63*(95). U.S. Environmental Protection Agency, Washington D.C.
EPA (1999). Federal Register Part II, 40 CFR 197, *Environmental Radiation Protection Standards for Yucca Mountain.* Nevada. Proposed Rule Federal Register 64/166, 46976–47016. U.S. Environmental Protection Agency (EPA).
EPA (2001a). *Improved Science-Based Environmental Stakeholder Process.* U.S. Environmental Protection Agency, EPA-SAB-EC-COM-01-006, www.epa.gov/sab
EPA (2001b). *IRIS Integrated Risk Information System.* U.S. Environmental Protection Agency (EPA). www.epa.gov/ngispgm3/iris/intro.htm

EPA (2001c). 40 CFR 197 *Public Health and Environmental Radiation Protection Standards for Yucca Mountain, Nevada.* Environmental Protection Agency, published 6 June 2001.

European Commission (1997). *European Commission Directive on Environmental Assessment:*85/337/EEC (as amended by 97/11/EC). European Commission, Brussels.

European Commission (2000). *The TRUSTNET Framework: a New Perspective on Risk Governance*. European Commission, Luxembourg, EUR 19150.

Fattah, A. (2000). *Safeguards for Geologic Repositories*, in IAEA (2000), op. cit.

Federline, M.V. (1993). *Views on Environmental Standards for Disposal of High-Level Wastes.* U.S. Nuclear Regulatory Commission Staff, Unpubl. Presentation to the National Academy of Sciences Committee on Technical Bases for Yucca Mountain Standards, Washington D.C.

Fritsche, A.F. (1992). *Wie gefährlich leben wir*. Verlag TUV, Rheinland.

Garwin, R.L., & Charpak, G. (2002). *Megatons and Megawatts: A Turning Point in the Nuclear Age*. Alfred A. Knopf, New York.

Ghiassi-nejad, M., Mortazavi, S.M.J., Cameron, J.R., Niroomand-rad, A. & Karam, P.A. (2002). Very high background radiation areas of Ramsar. Iran: preliminary biological studies. *Health Physics*, pp. 82, 87–93.

Goguel, J. (1987). *Le stockage des déchets radioactifs en formations géologique.* Rapport du groupe de travail présidé par le professeur Goguel. Ministére de l'industrie, Paris.

Goodwin, B.W., Cramer, J.J., & McConnell, D.B. (1989). *The Cigar Lake Uranium Deposit: An Analogue for Nuclear Fuel Waste Disposal in Natural Analogues in Performance Assessment for the Disposal of Radioactive Waste*. IAEA Technical Report 304.

Grupa, J.B., Dodd, D.H., Hoorelbecke, J.-M., Zuidema, P. et al. (2000). *Concerted action on the retrievability of long-lived radioactive waste in deep underground repositories*. EUR 19145 EN. European Commission, Brussels.

HSK & KSA (1993). *Protection Objectives for the Disposal of Radioactive Waste*. Guideline HSK-R21/e. Swiss Federal Nuclear Safety Inspectorate HSK and Federal Commission for the Safety of Nuclear Installations KSA. HSK, Wuerenlingen.

Hansson, S.O. (1998). *Setting the Limit: Occupational Health Standards and the Limits of Science*. New York: Oxford University Press.

Hedin, A. (1997). *Spent nuclear fuel — how dangerous is it?* Swedish Nuclear Fuel and Waste Management Company (SKB), Stockholm, Technical Report TR-97-13.

Holcomb, W.F., Clark, R.L., Dyer, R.S., & Galpin, F.L. (1988). USEPA *Radioactive Waste Disposal Standards:* Issued and Under Development. Nuclear and Chemical Waste Management 8/1, 3–12.

Hugi, M., Fritschi, M., Nold, A., Zuidema, P., & Smith, P. (1999). *Monitoring: Current status of the Swiss programme*. In: 5th International Workshop on Design and Construction of Final Repositories, Oxford, September 1999 (UK Nirex publisher).

Human Interference Task Force (1984). *Reducing the likelihood of future human activities that could affect Geologic High-Level Waste Repositories*. ONWI-537.

IAEA (1980). *Underground Disposal of Radioactive Wastes*. Proceedings of a Symposium in Otaniemi, July 1979. International Atomic Energy Agency (IAEA), Vienna.

IAEA (1981). *Safety Assessment for the Underground Disposal of Radioactive Wastes*. Safety Series 56, International Atomic Energy Agency (IAEA), Vienna.

IAEA (1983). *Criteria for Underground Disposal of Radioactive Wastes*. Safety Series 60. International Atomic Energy Agency (IAEA), Vienna.

IAEA (1985). *Performance Assessment for Underground Radioactive Waste Disposal Systems.* Safety Series 68, International Atomic Energy Agency (IAEA), Vienna, 198.

IAEA (1989). *Safety Principles and technical criteria for the Underground Disposal of High Level Radioactive Wastes.* Safety Series 99. International Atomic Energy Agency (IAEA), Vienna.

IAEA (1994a). *Siting of Geological Disposal Facilities.* Safety Series 111-G-4.1. International Atomic Energy Agency (IAEA), Vienna.

IAEA (1994b). *Safety indicators in different time frames for the safety assessment of underground radioactive waste repositories.* First report of the INWAC Subgroup on Principles and Criteria for Radioactive Waste Disposal. IAEATECDOC767. International Atomic Energy Agency (IAEA).

IAEA (1995a). *The Principles of Radioactive Waste Management.* Safety Series 111F. International Atomic Energy Agency (IAEA), Vienna.

IAEA (1995b). *Establishing a national system for radioactive waste management.* Safety Series 111-S-1. International Atomic Energy Agency (IAEA), Vienna.

IAEA (1995c). *Safety Fundamentals — The principles of radioactive waste management.* Safety Series 111-F, a publication within the RADWASS programme. International Atomic Energy Agency (IAEA), Vienna.

IAEA (1996a). *International Basic Safety Standards for Protection against Ionising Radiation and for the Safety of Radiation Sources.* Safety Series No. 115. International Atomic Energy Agency (IAEA), Vienna.

IAEA (1996b). *Issues in radioactive waste disposal — second report of the working group on principles and criteria for radioactive waste disposal.* IAEA-TECDOC-909. International Atomic Energy Agency (IAEA), Vienna.

IAEA (1997a). *The Joint Convention on the Safety of Spent Fuel Management and on the Safety of Radioactive Waste Management.* GOV/INF/821-GC(41)/INF/12. International Atomic Energy Agency (IAEA), Vienna.

IAEA (1997b). *Issues in radioactive waste disposal.* 2nd Report of the Working Group on Principles and Criteria for Radioactive Waste Disposal. Chapter 2, IAEA-TECDOC-909. International Atomic Energy Agency (IAEA), Vienna.

IAEA (1997c). *Optimisation of radiation protection — a review of its application to radioactive waste disposal.* In: Issues in radioactive waste disposal. IAEA-TECDOC-909, 16–21. International Atomic Energy Agency (IAEA), Vienna.

IAEA (1997d). *Regulatory decision making in the presence of uncertainty in the context of the disposal of long lived radioactive wastes.* Third report of the Working Group on Principles and Criteria for Radioactive Waste Disposal. IAEA-TECDOC-975. International Atomic Energy Agency (IAEA), Vienna.

IAEA (1997e). *Establishing a National System for Radioactive Waste Management.* Safety Series 111-S-1. International Atomic Energy Agency (IAEA), Vienna.

IAEA (1998a). *Technical, institutional and economic factors important for developing a multinational radioactive waste repository.* International Atomic Energy Agency (IAEA), Vienna.

IAEA (1998b). *Clearance of materials resulting from the use of radionuclides in medicine, industry and research.* IAEA-TECDOC-1000. International Atomic Energy Agency (IAEA), Vienna.

IAEA (1999b). *Protection of the environment from the effects of ionizing radiation.* IAEA-TECDOC-1091. International Atomic Energy Agency (IAEA), Vienna.

IAEA (1999c). *Maintenance of records for radioactive waste disposal.* IAEA-TECDOC-1097. International Atomic Energy Agency (IAEA), Vienna.

IAEA (1999d). *Application of radiological exclusion and exemption principles to sea disposal.* The concept of 'de minimis' for radioactive substances under the London Convention, 1972. IAEA-TECDOC-1068.

IAEA (2000). *Retrievability of high level Waste and spent Nuclear Fuel.* Proceedings of an international seminar in Saltsjöbaden, Sweden, IAEA-TECDOC-1187, Vienna, 2000.

IAEA (2001a). *Monitoring of Geological Repositories.* IAEA-TECDOC-1208. International Atomic Energy Agency (IAEA), Vienna.

IAEA (2001b). *Safety Requirements for the Geological Disposal of Radioactive Waste.* IAEA Safety Standards Series. International Atomic Energy Agency (IAEA), Vienna.

IAEA (2001c). *The Use of Reference Human Intrusion Scenarios in Safety Assessments of Radioactive Waste Disposal.* Draft.

IAEA (2002). *Issues Relating to Safety Standards on the Geological Disposal of Radioactive Waste.* Proceedings of a specialists meeting, Vienna, 18–22 June 2001. IAEA-TECDOC-1282. International Atomic Energy Agency (IAEA), Vienna.

IAEA (2003). *Developing and Implementing Multinational Repositories: Infrastructural Framework and Scenarios of Co-operation.* International Atomic Energy Agency (IAEA), Vienna, draft document.

ICA (1996). *Code of Ethics, adopted by the General Assembly in its XIIIrd session in Beijing (China) on 6 September 1996.* International Council on Archives, Available at: http://www.ica.org/c_ethics_e.html

ICRP (1977). *Recommendations of the International Commission on Radiological Protection.* International Commission on Radiological Protection ICRP Publication 26. Oxford: Pergamon Press.

ICRP (1979). *Limits for Intakes of Radionuclides by Workers*; a report of Committee 2 of the International Commission on Radiological Protection ICRP Publication 30. New York: Pergamon Press.

ICRP (1985). *Radiation Protection Principles for the Disposal of Solid Radioactive Waste.* International Commission on Radiological Protection ICRP Publication 46. Oxford: Pergamon Press.

ICRP (1991a). 1990 *Recommendations of the International Commission on Radiological Protection.* International Commission on Radiological Protection ICRP Publication 60. Volume 21, Numbers 1–3. Elmsford, New York: Pergamon Press.

ICRP (1991b). *Annual limits on intake of radionuclides by workers based on the 1990 Recommendations.* International Commission on Radiological Protection ICRP Publication 61. Pergamon Press, Oxford.

ICRP (1993). *Protection from potential exposure — a conceptual framework: radiation protection.* International Commission on Radiological Protection ICRP Publication 64. Pergamon Press, Oxford Annals of ICRP 23/1.

ICRP (1998a). *Radiation Protection Policy for the Disposal of Radioactive Waste.* International Commission on Radiological Protection ICRP Publication 77. Pergamon Press, Oxford.

ICRP (1998b). *Radiation Protection Recommendations as Applied to the Disposal of Long-Lived Solid Radioactive Waste.* International Commission on Radiological Protection ICRP Publication 81. Annals of ICRP 28/4.

ICRP (1999). *Protection of the Public in Situations of Prolonged Radiation Exposure.* The Application of the Commission's System of Radiological Protection to Controllable Radiation Exposure Due to Natural Sources and Long-lived Radioactive Residues.

International Commission on Radiological Protection ICRP Publication 82. Annals of the ICRP 29/1-2.

ICRP (2000a). *Radiation Protection Recommendations as Applied to the Disposal of Long-Lived Solid Radioactive Waste*. International Commission on Radiological Protection ICRP Publication 81. Annals of ICRP 28/4.

ICRP (2000b). *Protection of the Public in Situations of Prolonged Radiation Exposure*. The Application of the Commission's System of Radiological Protection to Controllable Radiation Exposure Due to Natural Sources and Long-lived Radioactive Residues. International Commission on Radiological Protection ICRP Publication 82. Annals of the ICRP 29/1-2.

ICRP (2001). A report on progress towards new recommendations: a communication from the International Commission on Radiological Protection. *J. Radiol. Prot.*, *21*, 113–123.

Imberger Elke (1997). *Die Bedeutung von technischen Normen für die Archivpraxis*, p. 115–122. In: Uhde, Karsten. Qualitätssicherung und Rationalisierungspotentiale in der Archivarbeit. Beiträge des 2. Archivwissenschaftlichen Kolloquiums der Archivschule Marburg.

INRA (2002). *Europeans and Radioactive Waste, Eurobarometer 56.2*. Prepared for European DG Energy and Transport, EC, DG Press and Communication, Brussels. 48 pps and Annex. http://europa.eu.intl/comm/energy/nuclear/pdf/EB56_radwaste_en_pdf

ISO 11108 (1996). *Information and documentation — Archival paper — Requirements for permanence and durability*. International Organisation for Standardisation, Geneva, Switzerland.

ISO 11798: *Anforderungen an die Dauerhaftigkeit von Schrift-, Druck- und Kopiermitteln*. International Organisation for Standardisation, Geneva, Switzerland.

ISO 11799: *Lagerungsbedingungen für Archiv- und Bibliotheksmaterial*. International Organisation for Standardisation, Geneva, Switzerland.

ISO 9706 (1994). *Paper for documents — Requirements for performance*, International Organisation for Standardisation, Geneva, Switzerland.

Jensen, M. (1993). *Conservation and retrieval of information: elements of a strategy to inform future societies about nuclear waste repositories*. Final report of the Nordic Nuclear Safety Research project KAN-1.3. Nordic Nuclear Safety Research.

Jauho, P., & Silvennoinen, P. (1980). *Warranty Obligations for the Management and Underground Disposal of Radioactive Waste*. International Atomic Energy Agency, Nuclear Energy Agency: Underground disposal of radioactive wastes: Proceedings of a symposium jointly organized by the IAEA and the OECD NEA and held at Otaniemi, Finland.

Kalbantner, P., & Sjöblom, R. (2000). *Techniques for freeing deposited canisters*. SLB Technical Report TR-00-15, SKB, Stockholm.

Kaplan, M.F. (1982). *Archaeological Data as a Basis for Repository Marker Design*. ONWI-354, Office of Nuclear Waste Isolation, Battelle Project Management Division, Columbus, Ohio.

Kaplan, M.F. (1986). Mankind's future: Using the past to protect the future. Archaeology and the disposal of highly radioactive wastes. *Interdisciplinary Science Reviews*, *11*(3), 257–268.

KASAM (1988). *Ethical Aspects on Nuclear Waste*. SKN Report 29, April 1988, SHN, Stockholm.

Kasperson (2000). *Risk and the Stakeholder Express*. RISK Newsletter, 4th Quarter 2000. http://www.sra.org/news.htm

Kellerer, A.M., & Nekolla, E.A. (2000). The LNT Controversy and the Concept of "Controllable Dose." *Health Physics*, *79*, 412.

Klett, R.D., & Gruebel, M.M., (1997). Development and Modification of Radioactive Waste Disposal Standards. In: *Radioactive Waste Management and Environmental Restoration* 21/2.

KSA (1992). *Sicherheitsprinzipien für die Entsorgung radioaktiver Abfälle*. Tagungsbericht der Klausurtagung 1992 des Ausschusses 'Strahlenschutz und Entsorgung'.

LaPointe, P., Wallman, P., Thomas, A., & Follin, S. (1997). *A Methodology to Estimate Earthquake Effects on Fractures Intersecting Canister Holes*. SKB Technical Report TR 97-07. SKB, Stockholm.

LaPointe, P., Cladouhos, T., & Follin, S. (1999). *Calculation of displacements on fractures intersecting canisters induced by earthquakes*. Aberg, Beberg and Ceberg examples. SKB Technical Report TR 99-03, SKB, Stockholm.

Larsson, C.-M., & Sundell-Bergman, S. (1999). Protection of the natural environment — philosophy and criteria. In: *Health and Environmental Criteria and Standards*, Stockholm Environment Institute, pp. 170–174.

Liljenzin, J.-O., & Rydberg, J. (1996). *Risks from nuclear waste*. Revised Edition. SKI Report 96:70, Swedish Nuclear Power Inspectorate, Stockholm.

Lindell, B. (1996). A history of radiation protection. *Radiation Protection Dosimetry*, *68*(1–2), 83–95.

Locke, P.A., Carney, M. C., Tran, N.L., Burke, T.A., & Melanson, M. (1998). *Chemical and Radiation Environmental Risk Management: Foundations, Common Themes, Similarities and Differences*. Paper for Workshop on Addressing the Similarities and Differences in Chemical and Radiation Risk Management, Annapolis 1998.

MacLachlan, A. (2003). ICRP, Bowing To Critics, ICPR Will keep Individual Dose Limits. *Nucleonics Week*, *44*(16), April.

Martin, J.E. (1988). Comparison of radioactive and chemical wastes. *Hazardous Waste and Hazardous Materials*, *5*(1).

Meadows, D.H., Meadows, D.L., Randers, J., & Behrens, W.W., III (1972). *The Limits to Growth. A Report to the Club of Rome*. New York: Universe Books.

McCombie, C., Zuidema, P., & McKinley, I.G (1991). *Sufficient Validation: The value of robustness in performance assessment and system design*. Validation of geosphere flow and transport models (Geoval), OECD/NEA, Paris, France.

McCombie, C. (2000). *Allocation of responsibilities for monitoring and retrieval activities*. In IAEA (2000a), p. 179; op. cit.

McCombie, C., & Stoll, R. (2002). *International and Regional Repositories: The Key Question*. Radwaste, May 2002, USA.

McKinley, I., & McCombie, C. (2002). Letters to the editor. *Nuclear Engineering International*, March 2002.

Miller, I., Black, J., McCombie, C., Pentz, D., & Zuidema, P. (1999). *High isolation sites for radioactive waste disposal: A fresh look at the challenge of locating safe sites for radioactive repositories*. Proceedings of Waste Management Conference (WM99), Tucson, Arizona, 1999.

Miller, P. (1998). *Quoted in Nuclear Fuel Waste Management and Disposal Concept*. Report of the Environmental Assessment Panel, Ministry of Public Works and Government Services, Canada.

Miller, W.M., Smith, G.A., Savage, D., Towler, P., & Wingefors, S. (1996). Natural radionuclide fluxes and their contribution to defining licensing criteria for deep geological repositories for radioactive waste. *Radiochimica Acta*, *74*, 289–295.

Miller, W.M., Alexander, W.R., Chapman, N.A., McKinley, I.G., & Smellie, J.A.T. (2001). *Geological Disposal of Radioactive Wastes and Natural Analogues*. Oxford: Pergamon Press, pp. 316.

Ministry of Industry and Trade (1991). *Statement of objectives to be applied to the safety of radioactive waste disposal in deep geological formations to ensure safety after the operating period of the repository*. Fundamental Safety Rule No. III.2.F, France. [Règle fondamentale de sûreté: Définition des objectifs à retenir dans les phases d'études et de travaux pour le stockage définitif des déchets radioactifs en formation géologique profonde afin d'assurer la sûreté après la période d'exploitation du stockage: RFS III.2.f. Ministère du Commerce et de l'Industrie, 1er juin 1991].

Mossman (2001). Deconstructing radiation hormesis. *Health Physics*, *80*(3), 263.

Nagra (1983). *Die Endlagerung schwach- und mittelradioaktiver Abfälle in der Schweiz: Evaluation der potentiellen Standortgebiete*. Nagra Technical Report NTB 83-15, Nagra, Wettingen.

Nagra (1985). *Nukleare Entsorgung Schweiz: Konzept und übersicht über das Projekt Gewähr 1985*. Nagra Gewähr Projekt NGB 85-01. Nagra, Wettingen.

Nagra (1988). *Sedimentsstudie — Zwischenbericht 1988: Möglichkeiten zur Endlagerung langlebiger radioaktiver Abfälle in den Sedimenten der Schweiz*. (Textband und Beilagenband). Nagra Technical Report NTB 88-25. Nagra, Wettingen.

Nagra (1989). *Sediment study — 1988: Disposal options for long-lived radioactive waste in Swiss sedimentary formations* — Executive summary. Nagra Technical Report NTB 88-25E. Nagra, Wettingen.

Nagra (1993a). *Vergleichende Beurteilung der Standorte Bois de la Glaive, Oberbauenstock, Piz Pian Grand und Wellenberg*. Nagra Technical Report NTB 93-02. Nagra, Wettingen.

Nagra (1993b). *Kristallin 1 — Safety assessment report*. Nagra Technical Report NTB 93-22. Nagra, Wettingen.

NAPA (1997). *Deciding for the future: balancing risks, costs and benefits fairly across generations*. Report for the USDOE by a panel of the U.S. National Academy of Public Administration, Washington D.C.

NAS (1957). *The Disposal of Radioactive Waste on Land*. Publication 519, Washington D.C., National Academy Press.

NEA (1982). Disposal of Radioactive Waste, an Overview of the Principles Involved. OECD/NEA, Paris.

NEA (1984a). *Geological Disposal of Radioactive Waste — An overview of the current status of understanding and development*. OECD/NEA, Paris.

NEA (1984b). *Long-Term Protection Objectives for Radioactive Waste Disposal. Expert Group Report*. OECD/NEA, Paris.

NEA (1985). *Technical Appraisal of the Current Situation in the Field of Radioactive Waste Management*. A Collective opinion by the Radioactive Waste Management Committee OECD/NEA, Paris.

NEA (1991). *Disposal of Radioactive Waste — Can Long-Term Safety Be Evaluated?* OECD/NEA, Paris.

NEA (1994). *Environmental and Ethical Aspects of Long-Lived Radioactive Waste Disposal*. Proceedings of an International Workshop OECD/NEA, Paris.

NEA (1995a). *The environmental and ethical basis of the geological disposal of long-lived radioactive wastes. A collective opinion of the Radioactive Waste Management*. Committee of the OECD Nuclear Energy Agency. OECD/NEA, Paris.

NEA (1995b). *Future human actions at disposal sites.* A report of the NEA Working Group on Assessment of Future Human Actions at Radioactive Waste Disposal Sites. OECD/NEA, Paris.

NEA (1997). *Regulating the Long-term Safety of Radioactive Waste Disposal. Proc.* of Conference in Cordoba, Spain, January 1997. Consejo de Seguridad Nuclear, Madrid, pp. 37.

NEA (1999a). *Confidence in the long-term safety of deep geological repositories: its development and communication.* OECD/NEA, Paris.

NEA (1999b). *Geological Disposal of Radioactive Waste. Review of Developments in the Last Decade.* OECD/NEA, Paris.

NEA (2000a). *Stakeholder Confidence and Radioactive Waste Disposal.* Workshop Proceedings, OECD/NEA, Paris, August, 2000.

NEA (2000b). *Developments in Radiation Health Science and their impact on Radiation Protection.* Committee on Radiation Protection and Public Health, OECD Nuclear Energy Agency, Paris.

NEA (2000c). *Features, Events and Processes (FEPs) for Geologic Disposal of radioactive Waste: An International Database.* OECD/NEA, Paris.

NEA (2000d). A Critical Review of the System of Radiation Protection: First Reflections of the NEA Committee on Radiation Protection and Public Health. OECD/NEA.

NEA (2001a). *Investing in trust: Nuclear Regulations and the Public.* Workshop proceedings. Paris, France 2000, OECD/NEA, Paris.

NEA (2001b). *Considering Reversibility and Retrievability in Geologic Disposal of Radioactive Waste.* Report NEA/RWM/RETREV (2001)2, OECD/NEA, Paris.

NEA (2001c). *The Way Forward: Modernisation of the system of Radiation Protection.* Draft report by CRPPH and EGRP.

NEA (2001d). *Reversibility and Retrievability in Geologic Disposal of Radioactive Waste.* OECD/NEA, Paris.

NEA (2002a). *Establishing and Communicating Confidence in the Safety of Deep Geologic Disposal.* OECD/NEA, Paris.

NEA (2002b). *The Way Forward in Radiological Protection.* OECD/NEA, Paris.

NEA (2003). *Radiological Protection of the Environment.* OECD/NEA, Paris.

NGL (1992). *The Nuclear Guardianship Library.* At www.nonukes.org/nlg.htm

Nolin, J. (1993). *Communicating with the future: implications for nuclear waste disposal.* Futures, Bd. 25, Nr. 7, S. 778–791.

NRC (1966). *Guidelines of U.S. National Academy Committee on Geological Aspects of Radioactive Waste.* Washington D.C., National Academy Press.

NRC (1990). *Rethinking High-Level Radioactive Waste Disposal.* National Research Council, Washington D.C., National Academy Press.

NRC (1992). *Radioactive Waste Repository Licensing: Synopsis of a Symposium.* Washington D.C., National Academy Press.

NRC (1995). *Technical Bases for Yucca Mountain Standards.* Board on Radioactive Waste Management, Commission on Geosciences, Environment and Resources & National Research Council (1995): Committee on Technical Bases for Yucca Mountain Standards. Washington D.C., National Academy Press.

NRC (1996). *Understanding Risk: Informing Decisions in a Democratic Society.* National Research Council, Washington D.C., National Academy Press.

NRC (2000a). *Research Needs in Subsurface Science*. National Research Council, Washington D.C., National Academy Press.
NRC (2000b). *Long-Term Institutional Management of US Department of Energy Legacy Waste Sites*. Washington D.C., National Academy Press.
NRC (2001a). *Disposition of High-Level Waste and Spent Nuclear Fuel.*— National Research Council, Washington D.C., National Academy Press.
NRC (2001b). *The Spent-Fuel Standard for Disposition of Excess Weapon Plutonium*. Washington D.C., National Academy Press.
NRC (2003). *One Step at a Time: The Staged Development of Geologic Repositories for High-Level Radioactive Waste*. Washington D.C., National Academy Press.
NRCP (2001). *Evaluation of the linear-nonthreshold dose-response model for ionising radiation*. NCRP Report No. 136. US National Council on Radiation Protection and Measurements, Maryland, USA, pp. 287.
NRPB (1992). *Board statement on radiological protection objectives for the land-based disposal of solid radioactive wastes. Advice report*. Documents of the NRPB 3/3. National Radiological Protection Board, Oxon.
NUMO (2002a). *Open Solicitation for Candidate Sites for Safe Disposal of High-Level Radioactive Waste*, December 2002. Nuclear Waste Management Organisation of Japan, Tokyo, Japan.
NUMO (2002b). *Siting Factors for the Selection of Preliminary Investigation Areas*. Nuclear Waste Management Organisation of Japan, Tokyo, Japan.
NWTRB (2001). *Report to the US Congress and the Secretary of Energy: Jan-Dec 2000*. U.S. Nuclear Waste Technical Review Board.
OECD (1997). OECD *Environmental Data Compendium, 1997*. Organisation for Economic Co-operation and Development, Paris.
Okrent, D. (1994). *On Intergenerational Equity and Policies to Guide the Regulation of Disposal of Wastes Posing Very Long Term Risks*. University of California Engineering Report No. UCLA ENG-22-94. School of Engineering and Applied Science, University of California, Los Angeles.
O'Neill, K. (2000). *Waste Trading Among Rich Nations*. Massachusetts, USA: MIT Press, Cambridge.
O'Neill, K. (2002). *Radioactive Trade: Globalizing the Nuclear Fuel Cycle*. SAIS Review XXII (1), pp. 157–168.
Overy, D.P., & Richardson, A.C.B. (1995). *Regulation of radionuclide and chemical carcinogens: current steps towards harmonization*. ELR News and Analysis 25 ELR 10657.
Paztor, S.B., & Hora, S.C. (1994). Lessons from the Vatican archives for repository recordkeeping. *Radwaste Magazine*, Bd. 1, Nr. 3, pp. 39–47.
Pellaud, B., McCombie, C. (2000). *International repositories for radioactive waste and spent nuclear fuel*. INMM, 41st Annual Meeting, New Orleans, 15–20 July 2000.
Pentreath, R.J. (1998). Radiological protection criteria for the natural environment. *Rad. Prot. Dosimetry*, *75*, 175–179.
Peterson P.E. (1998). *Post-closure repository safeguards — Comprehensive assessments of excavation methods*. Proceedings of the International High-Level Radioactive Waste Management '98, pp. 735–737.
Radiation Protection & Nuclear Safety Authorities in Denmark, Finland, Iceland, Norway & Sweden (1993). *Disposal of High-Level Radioactive Waste — Consideration of Some Basic Criteria*.

Raimbault, P., Valentin-Ranc, C. (1993). *How to mark repositories in geological formations.* SAFEWASTE 93: International Conference on Safe Management and Disposal of Nuclear Waste, Avignon, 13–18 June 1993, vol. 3, pp. 212–221.

RCEP (1998). *Setting Environmental Standards.* 21st Report of the Royal Commission on Environmental Pollution. Stationery Office, London.

Richardson, P.J. (1998). *A review of benefits offered to volunteer communities for siting nuclear waste facilities.* Swedish National Co-ordinator for Nuclear Waste Disposal, Stockholm.

Rochlin, G.I. (1997). Nuclear waste disposal — two social criteria. *Science, 195*, 23–31.

Rockwell (1997). What's wrong with being cautious? *Nuclear News*, June 1997. American Nuclear Society.

Roots, D. (1994). Radioactive Waste Disposal — Ethical and Environmental Considerations. a Canadian perspective. In NEA (1994), op. cit., p. 75.

RWMAC (2001). Advice to Ministers on the process for formulation of future policy for the long term management of UK solid radioactive Waste. Department for Environment Food and Rural Affairs, London: DEFRA Publications.

Savage, D. (ed). (1995) *The Scientific and Regulatory Basis for the Geological Disposal of Radioactive Waste.* John Wiley & Sons, Chichester.

Schrader-Frechette (1994). Risk and Ethics. In *Radiation and Society: Comprehending Radiation Risk*, IAEA, Vienna.

Sebeok, Th.A. (1984). *Communication measures to bridge ten millenia.* ONWI-532, Office of Nuclear Waste Isolation, Battelle Project Management Division, Columbus, Ohio.

Seitz, R. (1998). *Assessment of Health and Environmental Effects from Radioactive and Chemically Toxic Waste.* Proceedings of International Workshop on Comparative Evaluation of Environmental Toxicants Derived from Advanced Technologies, Chiba, Kodansha Scientific Ltd, Tokyo.

Selling, H.A. (2000). *Retrieval Disposal — Opposing Views on Ethics.* In IAEA (2000), pp. 137–144.

SKI, HSK & SSI (1990). *Regulatory guidance for radioactive waste disposal — an advisory document.* Swedish Nuclear Power Inspectorate (SKI), Swiss Nuclear Safety Inspectorate (HSK) and Swedish Radiation Protection Institute (SSI) Regulatory Working Group. SKI Report 90:15, Swedish Nuclear Power Inspectorate, Stockholm.

Slovic, P., Fischhoff, B., & Lichtenstein, S. (1985). *Characterizing Perceived Risk in Perilous Progress: Managing the Hazards of Technology.* Westview, Boulder, Colorado.

Smith, G.M., Little, R.H., & Watkins, B.M. (1999a). *Environmental risk assessment: its contribution to criteria development for HLW disposal.* In: 'Health and Environmental Criteria and Standards', Stockholm Environment Institute, pp. 175–178.

Smith, G.M., Apted, M.J., & Chapman, N.A. (1999b). *Human Intrusion and Effects on Multi-Barrier Disposal Systems.* Health and Environmental Criteria and Standards. Stockholm Environment Institute, pp. 285–288.

Söderberg, O. (2000). *Cost-related implications of retrieval: Who should pay? Who should assess the cost benefit?* In IAEA (2000a), p. 189.

SSI (1997). *Health, Environment and High Level Waste.* The Swedish Radiation Protection Institute's Proposed Regulations Concerning the Final Management of Spent Nuclear Fuel or Nuclear Waste. SSI Report 97-07. Swedish Radiation Protection Institute, Stockholm.

Stoll, R., McCombie, C. (2001). *The role of geologic disposal in preventing nuclear proliferation.* 9th International High-Level Radioactive Waste Management Conference, 2001, Las Vegas, USA.

Strebel, M. (1995). *Konservierung und Bestandeserhaltung von Schriftgut und Grafik: Ein Leitfaden für Archive.* Bibliotheken, Museen, Sammlungen. Schweiz. Verband für Konservierung und Restaurierung.

Stockholm Environment Institute (1999). *Health and Environmental Criteria and Standards.* Proceedings of the International Symposium on Radioactive Waste Disposal, SEI/USEPA/SSI, pp. 304.

STUK (2001). *Long-term safety of disposal of spent nuclear fuel.* Guide YVL 8.4. Finnish Radiation and Nuclear Safety Authority.

Tannenbaum, P.H. (1984). *Communication Across 300 Generations: Deterring Human Interference with Waste Deposit Sites.* ONWI-535.

Thunberg A.-M. (1999). *Retrievability in an ethical perspective,* in IAEA (2000), pp. 137 ff.

Tran, N.L., Locke, P.A., & Burke A. (2000). Chemical and Radiation Environmental Risk Management. Differences, Commonalties and Challenges. *Risk Analysis, 20*(2).

Trauth, K.M., Hora, S.C., & Guzowski R.V. (1993). *Expert judgement on markers to deter inadvertent human intrusion into the Waste Isolation Pilot Plant.* Sandia Report SAND92-1382. Sandia National Laboratories, Albuquerque.

UK Nirex (1999). *Radioactive Wastes in the UK: A Summary of the 1998 Inventory.* UK Nirex Ltd, N2-99-01.

UNCED (1992). *United Nations Conference on Environment and Development, Rio de Janeiro. Agenda 21.* (Full report and, in particular, Chapter 22: Safe and Environmentally Sound Management of Radioactive Wastes).

UNSCEAR (1993). *Sources and Effects of Ionizing Radiation.* UNScear 1993 Report to the General Assembly, with scientific annexes, No. E.94.IX.2. United Natios Scientific Committee on the Effects of Atomic Radiation, New York: United Nations.

UNSCEAR (1996). Effects of Radiation on the Environment. UN Scientific Committee on Effects of Atomic Radiation. 1996 Report to UN General Assembly: Annex 1, 7–86. New York: United Nations.

UNSCEAR (2000). *2000 Report to UN General Assembly, with Scientific Annexes.* United Nations Scientific Committee on the Effects of Atomic Radiation.

UNSCEAR (2000). *Sources and effects of ionizing radiation. United Nations Scientific Committee on the Effects of Atomic Radiation.* 2000 Report to UN General Assembly, Vol. 1: Sources and Vol. 2: Effects, with Scientific Annexes. New York: United Nations.

USC (1982). *Nuclear Waste Policy Act of 1982* (NWPA). 42 United States Congress, 42 U.S.C. 10101 et seq.

USC (1987). *Nuclear Waste Policy Amendments Act of 1987.* Public Law No. 100–203 101 Stat. 1330.

USNRC (1983a). 10 CFR 60, *Disposal of High-Level Wastes in Geological Repositories: Code of Federal Regulations.* U.S. Nuclear Regulatory Commission (NRC). Washington D.C.: U.S. Government Printing Office.

USNRC (1983b). 10 CFR 60.111(b): *Performance of the geological repository operations area through permanent closure:* Retrievability of waste. U.S. Nuclear Regulatory Commission (NRC).

USNRC (1999). *Proposed rule on Yucca Mountain* (Title 10, Part 63 of the Code of Federal Regulations: Disposal of High-Level Radioactive Wastes in a Proposed Geological Repository at Yucca Mountain, Nevada; Proposed Rule) in Federal Register (Volume 64, Number 34, February 22, 1999).

USNRC (2001). 10 CRF 63, *Disposal of High-Level Radioactive Wastes in a proposed geological repository at Yucca Mountain, Nevada: Final Version*. U.S. Nuclear Regulatory Commission, Federal Register, Nov. 2001.

Vieno, T., & Nordman, H. (1999). *Safety assessment of spent fuel disposal in Hästholmen, Kivetty, Olkiluoto and Romuvaara TILA-99*. Posiva Report 99-07.

Walker, J.S. (2000). *Permissible Dose. A History of Radiation Protection in the Twentieth Century*. University of California Press.

Weart, S.R. (1988). *Nuclear Fear. A History of Images*. USA: Harvard University Press.

Webb, G.A.M. (2001). From 'controllable dose' to the 'next recommendations'. Editorial. *J. Radiol. Prot.*, 21, June 2001.

Weber, H., & Dörr, M. (1997). *Digitisation as a method of preservation?* Final report of a working group of the German Research Association. European Commission on Preservation and Access, Amsterdam; Commission on Preservation and Access, Washington.

Webster, S. (2000). Stakeholders and the Public: Who are they? In NEA (2000a), p. 117.

Weitzberg, A. (1982). *Building on existing institutions to perpetuate knowledge of waste repositories*. ONWI-379.

Wingefors, S., Westerlind, M., & Gera, F. (1999). *The use of safety indicators in the assessment of radioactive waste disposal*. Health and Environmental Criteria and Standards. Stockholm Environment Institute, pp. 263–268.

WIPP (1992). *Land Withdrawal Act. Public Law;* 102–579.

World Commission on Environment and Development (1987). *Our Common Future*. New York: Oxford University Press.

Appendix 1

International Conventions and Agreements Concerning Deep Geological Disposal of Long-Lived Radioactive Wastes

Introduction

A limited number of international conventions and agreements affect deep geological repositories, although the global legal framework is expanding. Existing agreements can be grouped into those concerning:

- waste transport across international boundaries
- trans-boundary environmental impacts
- environmental information dissemination
- safe management of spent fuel and radioactive wastes
- regional agreements
- the marine environment

Whilst waste transport regulations are obviously important for the movement of wastes from one country to another, for example to regional or international waste disposal facilities, they are not discussed further here as they are not directly connected with repository safety and licensing issues.

Transboundary Environmental Impacts

The Espoo Convention on **Environmental Impact Assessment in a Transboundary Context** came into force in 1991. It is a regional convention affecting Europe, within the framework of the UN Economic Commission for Europe (UN-ECE). Signatory parties are committed to carrying out EIAs for any development that might have significant transboundary environmental effects, prior to a decision to authorise the development. The public, including people in neighbouring countries that may be affected, should have an opportunity to comment on the EIA. Article 3 of the Convention identifies radioactive waste disposal facilities as among those to which it will apply.

Environmental Information Dissemination

UN-ECE also developed the Aarhus **Convention on Access to Information, Public Participation in Decision-Making and Access to Justice in Environmental Matters** which came into force in 1998. As its name implies, signatories should provide access to information and allow for public participation in decision-making processes. Article 6 identifies installations designed for the final disposal of irradiated nuclear fuel or radioactive wastes as among those within its scope.

Safe Management of Spent Fuel

The year 1997 saw the adoption of the **Joint Convention on the Safety of Spent Fuel Management and on the Safety of Radioactive Waste Management**. The Convention came into force in June 2001, after it had been ratified by twenty five states, including fifteen with operational nuclear power plants. The objectives of this Convention are:

- to achieve and maintain a high level of safety worldwide in spent fuel and radioactive waste management, through the enhancement of national measures and international cooperation, including where appropriate, safety-related technical cooperation;
- to ensure that during all stages of spent fuel and radioactive waste management there are effective defenses against potential hazards so that individuals, society and the environment are protected from harmful effects of ionising radiation, now and in the future, in such a way that the needs and aspirations of the present generation are met without compromising the ability of future generations to meet their needs and aspirations;
- to prevent accidents with radiological consequences and to mitigate their consequences should they occur during any stage of spent fuel or radioactive waste management.

Chapter 3 of the convention, which is restricted to wastes from civilian nuclear programmes, covers in detail the activities that should be involved in the siting, design, safety, operation and institutional control of waste management facilities. These incorporate requirements of principle, including avoiding imposing undue burdens on future generations and avoiding actions that impose reasonably predictable impacts on future generations that are greater than those permitted today. There are also requirements to consult potentially affected parties in the vicinity of a facility and to provide safety information to the public. In relation to the previous discussion, the convention also requires contracting parties to ensure that facilities do not have unacceptable impacts on other contracting parties.

The convention requires contracting parties to establish and maintain a legislative and regulatory framework, and adequate financial resources and personnel to govern the safety of radioactive waste management, and to ensure that the regulatory function is independent. It notes that the prime responsibility for safety lies with the license holder for a facility, but, if none exists, the contracting party (State) is responsible.

With respect to post-closure safety indicators, Article 24 of the convention states that appropriate steps should be taken to ensure that discharges shall be limited:

- to keep exposure to radiation as low as reasonably achievable, economic and social factors being taken into account; and
- so that no individual shall be exposed, in normal situations, to radiation doses which exceed national prescriptions for dose limitation which have due regard to internationally endorsed standards on radiation protection.

The convention lies under the auspices of the IAEA, who acts as secretariat for meetings of the contracting parties. It can be seen that the principal articles of the convention are firmly based on or consistent with the IAEA Basic Safety Standards (IAEA, 1996a) and their 1995 Principles for Radioactive Waste Management (IAEA 1995b). As such, the convention is an important advance in placing such principles in a legal framework.

Regional (Non-Marine) Agreements

Many regional agreements concern potential marine pollution and these are discussed in the following section. The other principal major regional agreements are:

- The **Euratom Treaty** (1957), which requires European Union member states to provide the European Commission with such general data relating to any plan for the disposal of radioactive waste as will make it possible to determine whether its implementation is liable to result in radioactive contamination of the water, soil or airspace of another member state. The commission is required to deliver its opinion within six months, and this is likely to involve consultation with the member states.

  In November 2002, the EC proposed a Community approach to nuclear safety which included a "framework directive" to establish nuclear safety principles in the EU and regulate the management of decommissioning funds and a directive on the management and disposal of radioactive waste (CEC, 2002). Based for a large part on the IAEA "Joint Convention" (see above), the Directive on the Management of Spent Nuclear Fuel and Radioactive Waste provides that Member States should establish, according to a pre-set timetable, a strategy to deal with all categories of radioactive waste – focussing on geological disposal as the safest method, given our present state of knowledge. For high-level and long-lived waste — destined for geological disposal — Member States are required to have identified a repository site by 2008 and have it licensed by 2018. In the case of short-lived low- and intermediate-level waste — if this is to be disposed of separately from high-level waste — authorisation for operation of the disposal facility should be granted no later than 2013. Repositories maybe shared between countries, provided that exports of radioactive waste or spent fuel to other Member States are fully in compliance with existing EU legislation. At the time of writing this book, the directive was still under discussion between EU Member States.

- The **Antarctic Treaty** (1959), which specifically prohibits the disposal of radioactive wastes in Antarctica.

The Marine Environment

Many international conventions that could affect deep disposal of long-lived wastes are, in fact, concerned with marine pollution. These conventions not only concern direct disposal of wastes to the sea, but could also affect deep geological repositories located on the coast, beneath the seabed off the coast, or indeed, any deep repository inland that might be construed as eventually releasing radionuclides to the seas. The international legal and political position is by no means clear-cut, and is currently evolving. Much depends on the extent to which a repository might be regarded (both nationally and internationally, especially on a regional basis) as a potential contributor to marine pollution, and on the interpretation of a number of conventions, declarations and agreements concerned with waste "dumping", with protection of the marine environment and with the management of coastal zones. These are discussed below.

Essentially three areas are of relevance:

- the political definitions of maritime zones, for a repository situated offshore (that is, beneath the seabed at some distance from the coast);
- agreements and conventions affecting waste disposal on or under the seabed, which might affect an offshore repository;
- agreements and conventions concerning pollution of the marine environment, which might be considered to affect both an offshore repository or one located inland but at the coast.

The first of these is dealt with briefly below, while the second two areas are dealt with together in detail, as they are closely related and can most conveniently be considered under the headings of *global* and *regional* agreements.

Definition of Maritime Zones

International agreement on the definition of maritime zones and activities which could be carried out within them is one of the principal achievements of UNCLOS, the United Nations Convention on the Law of the Sea (1982), which is itself based on agreements on Territorial Seas (etc) dating back to 1958. UNCLOS defines four maritime areas which can be summarised in brief as:

Territorial Seas and the Contiguous Zone The Territorial Sea is a region of 12 nautical miles (nm) from an agreed baseline (normally low water mark, but with special provisions for baselines in archipelagic states). In this region, states have full sovereignty over the sea and the seabed. The Contiguous Zone (a region a further 24 nm beyond the Territorial Sea) is the region in which states can protect their interests and, in the present context, is of no direct relevance.

The Exclusive Economic Zone (EEZ) The EEZ extends beyond the Territorial Seas, but is measured from the same baseline out to a distance of 200 nm. In the present context, the relevance of the EEZ lies in a state's sovereign rights "*for the purpose of exploring and exploiting, conserving and managing the natural resources, whether living or non-living, of the waters superjacent to the seabed and of the seabed and its subsoil, and with regard to other activities for the economic exploitation and exploration of the zone, such as the production of energy from the water, currents and winds*" and in a state's jurisdiction with regard to "*the establishment and use of artificial islands, installations and structures and... the protection and preservation of the marine environment...*" The construction of artificial islands is subject to various strictures about location with respect to shipping lanes, removing them after use, etc.

The Continental Shelf This is "the sea-bed and subsoil of the submarine areas that extend beyond its territorial sea throughout the natural prolongation of its land territory to the outer edge of the continental margin, or to a distance of 200 nm from the baselines from which the breadth of the territorial sea is measured where the outer edge of the continental margin does not extend up to that distance". It thus covers much the same region as the EEZ, but it can extend (and be claimable) further, to cover any areas of physical continental shelf beyond 200 nm.

The Area This is essentially the sea floor beyond the outer edge of the continental margin. This underlies almost the same region as is covered by the **High Seas**, which lie beyond the EEZ (rather than the edge of the continental margin) and are regarded as the "common heritage of mankind".

In terms of repository development, there would thus appear to be no jurisdictional issue within this part of UNCLOS in constructing a repository beneath the seabed within the Territorial Seas or the EEZ, or from an artificial island within any country's EEZ or Continental Shelf region, if construction of a repository can be regarded as exploitation of a natural resource (the rock mass). However, other conventions and agreements also influence this matter, and are more restrictive, as described in the following sections.

Waste Disposal in the Seabed and Marine Pollution

Essentially, six existing conventions and agreements have global influence. World-wide, there are also agreements that are of a *regional* nature (e.g. only affecting the Pacific region, the Baltic, the Black Sea, etc), reflecting the direction of currently more localised initiatives on potential marine pollution. With the exception of OSPAR (see below), the latter are not discussed in this book. Those with a global bearing are:

- Agenda 21 (Chapter 17: Protection of the Oceans): the United Nations Conference on Environment and Development (UNCED; Rio de Janeiro, 1992);
- UNCLOS: the United Nations Convention on the Law of the Sea (1982);
- UNEP-GPA: the United Nations Environment Programme (UNEP) Global Programme of Action (GPA) for the Protection of the Marine Environment from Land-based Activities (derived from the 1985 Montreal Guidelines): most directly the:
- UNEP-GPA Washington Declaration (1995) on the Protection of the Marine Environment from Land-based Activities;
- London Convention (1972) and LC Protocol (1996): the Convention on the Prevention of Marine Pollution by Dumping of Wastes and Other Matter (LC72);
- UNESCO-IOC Ocean Charter: the International Oceanographic Commission "Ocean Charter" produced during the 1998 "International Year of the Ocean";

Although all of the regimes listed above are concerned with protection of the marine environment, it is important to appreciate at the outset that there is no global agreement that deals specifically with the prevention and control of marine pollution from land-based activities, in the way that the London Convention controls dumping at sea.

Those conventions or agreements which are of a regional nature, but still relevant because they are indicators of the direction of current international policy developments are:

- OSPAR Convention (1992): convention of the Oslo and Paris Commission for the Protection of the Marine Environment of the North-East Atlantic (entered force in 1998), and more specifically the:
- OSPAR (Sintra Meeting) Strategy with regard to Radioactive Substances (1998).

Global Conventions and Declarations

AGENDA 21

Although it has no legal effect, the UNCED 1992 Rio de Janeiro consensus document (called Agenda 21) essentially set the stage for much of the current direction of national and global environmental strategic thinking, and the terminology and the concepts embodied in it would need to be considered when making a safety case for any repository, coastal or otherwise. Specifically, Agenda 21 enshrined "sustainability" and the "precautionary principle" firmly into international environmental policies.

The most direct reference with respect to a coastal repository is to be found in Chapter 22, where Paragraph 22.5 states:

> States, in cooperation with relevant international organisations, where appropriate, should:...
>
> (c) Not promote or allow the storage or disposal of high-level, intermediate-level and low-level radioactive wastes near the marine environment unless they determine that scientific evidence, consistent with the applicable internationally agreed principles and guidelines, shows that such storage or disposal poses no unacceptable risk to people and the marine environment or does not interfere with other legitimate uses of the sea, making, in the process of consideration, appropriate use of the concept of the precautionary approach;

Chapter 17 is entitled "Protection of the Oceans, All Kinds of Seas, Including Enclosed and Semi-Enclosed Seas, and Coastal Areas and the Protection, Rational Use and Development of their Living Resources". This Chapter makes reference in several places to the UN Convention on the Law of the Sea, which is discussed later. The following paragraphs of Chapter 17 are of particular relevance:

> 17.21. A precautionary and anticipatory rather than a reactive approach is necessary to prevent the degradation of the marine environment. This requires, inter alia, the adoption of precautionary measures, environmental impact assessments, clean production techniques, recycling, waste audits and minimisation, construction and/or improvement of sewage treatment facilities, quality management criteria for the proper handling of hazardous substances, and a comprehensive approach to damaging impacts from air, land and water. Any management framework must include the improvement of coastal human settlements and the integrated management and development of coastal areas.

It can be seen that a coastal repository development programme would need to take particular note of Paragraph 22.5 (c), which raises the matter of the precautionary principle that has been used frequently by groups opposed to all types of geological disposal on the grounds that the impacts are poorly understood, so we should wait until they are better understood to avoid causing irreparable harm to the environment. In this case, the argument would concern potential releases to, and harm to, the marine environment. In this context, it could be argued that application of "*scientific evidence, consistent with the applicable internationally agreed principles and guidelines*" in the form of a comprehensive performance assessment and EIS would scope any potential harm adequately and demonstrate that risks were acceptable.

UNCLOS

The text of the UN Convention on the Law of the Sea dates back to 1982, although the concept originated in 1958. The relevant sections of this extensive document are Part VI, which defines the sea areas around maritime nations (as discussed above), and Part XII which concerns protection and preservation of the marine environment. As noted above, Part VI acknowledges the right of any country to drill into the seabed and exploit the resources within the 200 nm EEZ. The relevant text of Part XII includes the following:

Article 194

Measures to prevent, reduce and control pollution of the marine environment

1. States shall take, individually or jointly as appropriate, all measures consistent with this Convention that are necessary to prevent, reduce and control pollution of the marine environment from any source, using for this purpose the best practicable means at their disposal and in accordance with their capabilities, and they shall endeavour to harmonise their policies in this connection.
2. States shall take all measures necessary to ensure that activities under their jurisdiction or control are so conducted as not to cause damage by pollution to other States and their environment, and that pollution arising from incidents or activities under their jurisdiction or control does not spread beyond the areas where they exercise sovereign rights in accordance with this Convention.
3. The measures taken pursuant to this Part shall deal with all sources of pollution of the marine environment. These measures shall include, inter alia, those designed to minimise to the fullest possible extent:

(a) the release of toxic, harmful or noxious substances, especially those which are persistent, from land-based sources, from or through the atmosphere or by dumping;

It can be seen that this aspect of UNCLOS has been incorporated conceptually in Agenda 21. Paragraph 2 (above) clearly has implications if neighbouring countries feel that they may be affected by potentially contaminating activities, as did Ireland when faced with the UK Nirex deep repository proposals for Sellafield. In this context, any state in the region of a country siting a repository on the coast might consider that it has a legitimate concern.

In addition, **Article 197** states that:

States shall co-operate on a global basis and, as appropriate, on a regional basis, directly or through competent international organizations, in formulating and elaborating international rules, standards and recommended practices and procedures consistent with this Convention, for the protection and preservation of the marine environment, taking into account characteristic regional features.

This (along with Article 207: see below) led eventually to the current UNEP-GPA which is endeavouring to move towards such standards and recommended practices (see below).

Articles 207 and 208 are specifically concerned with marine pollution from land-based sources and sea-bed activities:

Article 207

Pollution from land-based sources

1. States shall adopt laws and regulations to prevent, reduce and control pollution of the marine environment from land-based sources, including rivers, estuaries, pipelines and outfall structures, taking into account internationally agreed rules, standards and recommended practices and procedures.

Article 208

Pollution from sea-bed activities subject to national jurisdiction

1. Coastal States shall adopt laws and regulations to prevent, reduce and control pollution of the marine environment arising from or in connection with sea-bed activities subject to their jurisdiction and from artificial islands, installations and structures under their jurisdiction, pursuant to articles 60 and 80.

Article 207 could clearly be argued to apply to a coastal repository (as well as to an inland repository discharging to the sea), while Article 208 could be argued to apply to a sub-seabed repository or one accessed from an artificial island, and both refer to the need to apply international and regional regulations to any potentially polluting activity.

Finally, Articles 204 and 205 place an onus on the Governments of states giving rise to pollution to evaluate and monitor potential environmental impacts and make the results of such evaluations openly available to other countries, via competent international organisations:

Article 205

Publication of reports

States shall publish reports of the results obtained pursuant to article 204 or provide such reports at appropriate intervals to the competent international organizations, which should make them available to all States.

Article 206

Assessment of potential effects of activities

When States have reasonable grounds for believing that planned activities under their jurisdiction or control may cause substantial pollution of or significant and harmful changes to the marine environment, they shall, as far as practicable, assess the potential effects of such activities on the marine environment and shall communicate reports of the results of such assessments in the manner provided in article 205.

UNEP-GPA

In 1982, the United Nations Environment Programme (UNEP) took the initiative to develop advice to Governments on addressing impacts on the marine environment from land-based activities. This initiative resulted in the preparation of the **Montreal Guidelines** for the Protection of the Marine Environment Against Pollution from Land-based Sources in 1985. Following the successful international adoption of Agenda 21 in 1992, which now provided a stronger general environmental framework, the concept was revisited by UNEP in 1995. They developed a Global Programme of Action (GPA) for the Protection of the Marine Environment from Land-based Activities whose aims are contained in the so-called **Washington Declaration**.

The Declaration states that:

> The Global Programme of Action aims at preventing the degradation of the marine environment from land-based activities by facilitating the realization of the duty of States to preserve and protect the marine environment. It is designed to assist States in taking actions individually or jointly within their respective policies, priorities and resources, which will lead to the prevention, reduction, control and/or elimination of the degradation of the marine environment, as well as to its recovery from the impacts of land-based activities.

Again, The GPA has no legal status, but reflects international thinking and aspirations, being intended more as a source of conceptual and practical guidance for national and regional authorities.

Section V(C) recommends approaches for managing marine pollution from radioactive substances, with paragraph 109 saying:

> The objective/proposed target is to reduce and/or eliminate emissions and discharges of radioactive substances in order to prevent, reduce and eliminate pollution of the marine and coastal environment by human-enhanced levels of radioactive substances.

The paragraphs that follow in the Declaration recommend actions that should be taken at the national, regional and international level. They reiterate, almost verbatim, paragraph 22.5c of Agenda 21 on siting repositories near the marine environment, except the scope is broadened to "marine *and coastal* environments". They are cited here at some length as they are effectively the most recent and relevant international statement of the trend of environmental thinking:

> **(a) National actions, policies and measures**
>
> 110. Actions, policies and measures of States within their national capacities should include:
>
> (a) Promotion of policies and practical measures including setting targets and timetables to minimize and limit the generation of radioactive wastes and provide for their safe processing, storage, conditioning, transportation and disposal;
> (b) Ensuring the safe storage, transportation and disposal of radioactive wastes, as well as spent radiation sources and spent fuel from nuclear reactors destined for final disposal, in accordance with international regulations or guidelines;

111. States should:

(a) Not promote or allow the storage or disposal of high-level, intermediate-level and low-level radioactive wastes near the marine and coastal environment unless they determine that scientific evidence, consistent with the applicable internationally agreed principles and guidelines, shows that such storage or disposal poses no unacceptable risk to people and the marine and coastal environment or does not interfere with other legitimate uses of the sea, making, in the process of consideration, appropriate use of the concept of the precautionary approach;

(d) Make available information on the characteristics of terrestrial dump sites in coastal areas through, and consistent with, agreed regional and international reporting procedures. The information should include the magnitude, types of materials, characteristics of storage and status of the dump sites.

(b) Regional actions

112. Relevant regional organizations, in accordance with regional needs and capacities, should ensure:

(a) Monitoring of radioactivity in their regions and identification of any problem areas;

(b) The establishment of criteria for assessing and/or reporting on the use in their region of best available techniques to prevent and eliminate pollution by inputs of radioactive substances;

(c) The preparation of comprehensive environmental assessments of the effect on the marine and coastal environment of historical discharges and current discharges of radioactive substances.

(c) International actions

113. International actions should include:

(a) Support for efforts under the auspices of IAEA to develop and promulgate radioactive waste management safety standards, guidelines or codes of practice, including work being undertaken towards an international convention on the safety of radioactive waste management, in order to provide an internationally accepted basis for the safe and environmentally sound management and disposal of radioactive wastes. This work should take account of the application of best available techniques and best environmental practice for all nuclear applications not currently covered by internationally binding agreements making such provisions;

(b) Cooperation with countries in need of assistance, through financial, technical and scientific support, in ensuring environmentally sound management and storage of radioactive materials as well as supporting environmental restoration efforts;

Clearly, many of the above recommendations are aimed at management of radioactive wastes and potential pollution sources in general. Nevertheless, some of them would have a bearing on how a proposal for a deep repository near the coast might be received, both nationally and regionally. For example, paragraph 110d suggests "*reduction and/or elimination*" of inputs of radioactive substances to the marine environment. This approach has certainly been adopted in the 1998 Sintra agreement of OSPAR.

Paragraph 111d could be taken to require states to make information on a coastal repository widely available to neighbouring countries and this could be at the planning or design stage, and paragraph 112 goes on to propose setting up regional criteria for assessing techniques for preventing or eliminating marine inputs.

London Convention

The London Convention of 1972, sometimes called LC72 or the London Dumping Convention, is specifically concerned with the dumping or discharge of wastes directly to the marine environment (e.g. from ships) and thus appears, in principle, only to be of marginal concern. However, a Protocol to the Convention was agreed in 1996 which amended the definition of both dumping and of the sea. The Protocol is not currently in force, as it has first to be ratified by a minimum number of 26 states, but when it comes into force, it will supersede the Convention for those states that sign up to it.

The Convention (and the Protocol) effectively prohibit the dumping or radioactive wastes, as they are not among the list of wastes or other matter that may be considered for dumping. The Protocol goes further, saying that materials containing levels of radioactivity greater than *de minimis* (exempt) concentrations, as defined by the IAEA,[24] shall not be considered eligible for dumping. It then makes the provision that within 25 years (from February 1994), and at each 25 year interval afterwards, contracting parties to the Protocol shall complete a scientific study relating to all radioactive wastes (other than HLW) and review the prohibition on dumping them.

The relevance to the present discussion comes in Article 1 of the Protocol, which now includes within the meaning of dumping "any storage of wastes or other matter in the seabed and the subsoil thereof from vessels, aircraft, platforms or other man-made structures at sea". The definition of "sea" has also been amended to include the seabed and the subsoil thereof, with the exclusion, however, of sub-seabed repositories accessed only from the land.

Thus, the new Protocol would seem to prevent the development of a deep repository accessed from a platform, but *not* one accessed from land, and presumably also prohibits a repository accessed from an artificial island (a "man-made structure at sea").

---

[24]The IAEA issued its first advice to the London Convention on exclusion and exemption principles in 1999 (IAEA, 1999d).

UNESCO-IOC Ocean Charter

1998 was declared as the "International Year of the Ocean" by the UN General Assembly, and many countries signed the Ocean Charter, a short and non-specific declaration to the effect that the oceans are a vital resource which must be used sustainably and protected internationally. Whilst this has no legal standing, it is a further indicator of the swell of opinion towards protecting the oceans, and something which could be raised in objection to a proposal for a coastal or sea-bed repository.

Regional Conventions and Declarations

OSPAR Convention

The 1992 convention of the Oslo and Paris Commission for the Protection of the Marine Environment of the North-East Atlantic (which entered into force in 1998), and more specifically the OSPAR Strategy with Regard to Radioactive Substances (agreed at the Sintra, Portugal, meeting in 1998) have relevance to a coastal repository more because they represent a trend in current environmental controls on marine releases of radioactivity, albeit at a regional level (of the NE Atlantic area, in this case).

As with Agenda 21 and the Washington Declaration, OSPAR also requires the application of the precautionary principle and recognises that it may be desirable to adopt, on the regional level, more stringent measures when considering the prevention and elimination of marine pollution than are provided for in international conventions or agreements with a global scope. In this latter sense, OSPAR is breaking into new ground.

The OSPAR Convention defines land-based sources of pollution as:

> point and diffuse sources on land from which substances or energy reach the maritime area by water, through the air, or directly from the coast. It includes sources associated with any deliberate disposal under the sea-bed made accessible from land by tunnel, pipeline or other means and sources associated with man-made structures placed in the maritime area under the jurisdiction of a Contracting Party, other than for the purpose of offshore activities.

Consequently, OSPAR would appear to affect a deep coastal repository.

The parts of the Convention concerned with pollution from land-based sources are contained in Article 3 and Annex I. Article 3 states that the Contracting Parties shall take, individually and jointly, all possible steps *to prevent and eliminate* pollution from land-based sources. However, the details presented in Annex I and which would have an impact on a coastal repository contain nothing especially novel:

However, the Annex presages potential changes in the future in Article 3 when it states "it shall, inter alia, be the duty of the Commission to draw up...plans for the reduction and phasing out of substances that are toxic, persistent and liable to bioaccumulate arising from land-based sources..."

In this context, the Sintra agreement in 1998 laid out the OSPAR Strategy with regard to Radioactive Substances, which is considered to have potentially far-reaching effects for the nuclear industry, as well as implications for radioactive waste repositories. The stated objective of this strategy is as follows:

> In accordance with the general objective, the objective of the Commission with regard to radioactive substances, including waste, is to prevent pollution of the maritime area from ionising radiation through progressive and substantial reductions of discharges, emissions and losses of radioactive substances, with the ultimate aim of concentrations in the environment near background values for naturally occurring radioactive substances and close to zero for artificial radioactive substances. In achieving this objective, the following issues should, inter alia, be taken into account:
>
> a. legitimate uses of the sea;
> b. technical feasibility;
> c. radiological impacts on man and biota.

The time frame for achieving this objective is as follows:

- By the year 2000: the Commission will, for the whole maritime area, work towards achieving further substantial reductions or elimination of discharges, emissions and losses of radioactive substances;
- By the year 2020: the Commission will ensure that discharges, emissions and losses of radioactive substances are reduced to levels where the additional concentrations in the marine environment above historic levels, resulting from such discharges, emissions and losses, are close to zero.

It is clear that the Sintra wording will need to be developed and interpreted before any real application. The definition of the terms "historic levels" and "close to zero" in the OSPAR strategy are of particular concern. One approach that may be explored is to evaluate the general Sintra statements on radionuclide concentrations in terms of the doses which might arise from given concentrations in different environmental compartments. Concentrations might be considered to be low enough to meet the aims of the declaration if they only gave rise to doses which would normally be below levels of concern: generally, in the few, or few tens of microsievert range.

For example, in 2002, the UK Government proposed a strategy whereby impacts from discharges after 2020 should not exceed 0.02 mSv/a to a member of a critical group of the general public. At the time of writing, Belgium, France, Spain and the UK were asking for a different "baseline" period to represent "historic levels", proposing the 1993–1997 (pre-Sintra) period rather than the 1996–2000 period suggested by the OSPAR Radioactive Substances Committee and when discharges from the UK and France and already been substantially reduced. The next ministerial meeting to follow the Sintra meeting took place in June 2003 in Bremen.

The issue of "baselines" of time and concentrations was still very much to the fore but the 2020 target date was still regarded as achievable.

Discussion

Perhaps the most significant international agreements and conventions that impact on principles and standards for deep geological disposal concern the possibility of pollution of the seas. The last decade has seen a considerable acceleration in interest internationally in preventing marine pollution and in protecting the seas. 60% of the world's population lives within 60 km of the coastline and the coastal zone is the focus of much of the world's economic activity. The GPA and the Year of the Ocean are symptomatic of the increasing interest in protecting and, presumably, eventually regulating activities in the coastal areas of the world. Several countries are now independently developing their own specifically marine environmental policies. For example, Australia now has an "Oceans Policy" and the UK has a "Cleaner Seas" initiative.

"Integrated Coastal Management" (ICM) is now an important part of many national environmental initiatives and the significance of coastal management worldwide is testified by the interest taken by all the international agencies. For example, the International Maritime Organisation (IMO), the UN Food and Agriculture Organisation (FAO), the IAEA, UNEP, the UNESCO International Oceanographic Commission and the World Health Organisation all participate in GESAMP, the Joint Group of Experts on the Scientific Aspects of Marine Environmental Protection. GESAMP has a task force on ICM which is intended to promote sustainable development of coastal regions.

Despite all this activity, there are currently no international agreements which control activities in the coastal region that might pollute the seas, and the sum of all the initiatives mentioned in the previous paragraphs is indicative only of a growing environmental concern. A direct example of this lack of enforceable agreements is the experience of the Irish Government, over several years, in trying to have a say in the control of radioactive discharges into the Irish Sea from the Sellafield reprocessing plant in the UK. Several years ago, the Irish Government suggested that they might seek an amendment to the EURATOM treaties to take account of non-nuclear jurisdictions sharing land or maritime borders with countries that operate nuclear power and reprocessing plants, to include stringent regulations for decommissioning such facilities, and prohibiting underground waste repositories which might pollute groundwater and marine resources. They have also suggested an Irish Sea Inter-Governmental Conference to make rules on discharges to the Irish Sea. To some extent, these developments were superseded by the 1998 Sintra OSPAR agreement which has moved the UK substantially in the direction in which Ireland was pushing. Sintra, along with the thrust of GPA, suggest that it is feasible that conventions and agreements may enter force over the next two decades which would translate today's environmental concerns into mandatory practice in many countries. How might this affect a deep repository development programme, especially one in or near a coastal area? In this context, and summarising what the existing agreements discussed earlier have to say on the matter, the following conclusions can be drawn:

Neither Agenda 21 nor UNCLOS preclude the development of a coastal or offshore sub-seabed repository. Indeed, UNCLOS appears to permit the construction of artificial islands many kilometres from the coast in the EEZ or continental shelf area, drilling and exploitation of seabed resources, which could be argued to include sub-seabed space. The thrust of these two conventions seems to be that, if such a policy is pursued, then a proper environmental impact assessment must be carried out, the project should comply with international standards for radioactive waste repositories and the programme should be transparent to other interested (e.g. neighbouring) countries. None of these stipulations should pose any unusual problems, although there may be more pressure to have programmes internationally reviewed than would be the case for an inland repository. Similarly, none of the other environmental pollution conventions and declarations (with the partial exception of LC 96: see below) would prevent development of a repository, but all would aim at using the best possible technology to reduce releases and at performing the highest level of environmental assessment using the best science. It is worth noting that there is a tendency for all the documents to be phrased more in the context of existing practices and sources rather than entirely new ones, and the possible reactions to proposing something totally new for a region need to be considered.

Notwithstanding this apparent window through which to proceed, the same documents also provide potential reasons to hinder a repository programme. First, some parties would argue that application of the precautionary principle, demanded by Agenda 21, militates against any form of radioactive waste disposal. The counter argument is that the *effects* of releases of radioactivity are well understood (we are not unleashing an unknown contaminant into the environment), although the magnitude, time and location of such releases is more difficult to evaluate. A second reason for caution lies in the UNCLOS Article 194 phrase "*to minimise to the fullest possible extent, the release of toxic, harmful or noxious substances*". It might be argued that any disposal concept which envisages slow releases of activity is not minimising them, compared to other potential disposal options. Again, counter arguments can be found in the sphere of optimisation and ALARA.

UNCLOS contains requirements to reduce or prevent contamination from artificial islands, although this presumably did not have their possible use as access points for a sub-seabed repository in mind. Critically, LC Protocol 96 would preclude a repository being constructed with access from a platform (and, apparently, an artificial island), although allowing one accessed from the land (which is not included in their definition of seabed). There is some ambiguity here which would need to be resolved, particularly as these two documents are the most powerful agreements to hand (although LC 96 is not yet activated).

The OSPAR strategy for radioactive substances might now be seen to be leading the way in terms of possible future global conventions. The aim of achieving near-background environmental concentrations of naturally occurring radionuclides is unlikely to be onerous to a repository developer and is indeed, already part of several national regulations for long-term (e.g. after about 10 to 100 ka) releases from long-lived waste repositories. The fact that there might be some such requirement for concentrations in seawater and seabed sediments ought to be being considered in any new standards being developed nationally, along with the related issue of the timescale to be applied. On the other hand, the OSPAR objective for "artificial radionuclides" of close to zero concentrations is still being interpreted. The application of *de minimis* concentration values (as envisaged in LC 96) might be seen to meet this objective, but the question then arises as to the definition of the size, location and content of environmental compartments to which they are applied. As noted above, this matter will be the subject of discussion and development within OSPAR member countries over the next few years.

The sum total of all these considerations would appear to be that a repository could be built under the seabed, provided it is accessed from the coast, or it could be built inland at the coast, but that releases from it may soon (if the OSPAR lead is followed) have to be demonstrated to be below *de minimis* levels, close to background, or close to zero, depending on radionuclide and on interpretation. An artificial island seems to be about to be ruled out by LC 96, when it comes into force. For a repository accessed from the land, there appears to be no reason why this cannot be outside the Territorial Sea (beyond 12 nautical miles) if climate evolution considerations suggest that to be appropriate.

All of the following discussion must then be re-evaluated in the light of the likely long-term behaviour of a repository. Only certain locations would give rise to direct releases to the sea at all times over the next million years. Others, in common with inland repositories, would discharge indirectly to the sea, via lakes and rivers at certain periods in the future. But minute quantities of radionuclide from all repositories, no matter where located, will ultimately find their way to the sea to some extent. This was an issue which certainly caused confusion for Nirex in the UK, at the Sellafield public inquiry, when talking about releases to the Irish Sea, which is an ephemeral feature when looked at over the past, and the next million years. None of the marine environmental conventions discussed above is concerned with such long time periods. The way in which any of them might be cited as a reason not to proceed with a repository, as it inevitably pollutes the sea, might appear to be an entirely philosophical matter, but it is an issue which the implementor must be prepared to debate.

Appendix 2

Development of Radiation Protection Standards for Geological Disposal of Radioactive Wastes in the USA

In this Appendix we examine in some detail the development of radiation protection standards for disposal of HLW, spent nuclear fuel and long-lived wastes in the USA. The reasons for the close examination are not only that the USA has licensed for operation the world's first custom-built deep geological repository (the WIPP facility in New Mexico) and is preparing a license application for disposal of spent fuel (at Yucca Mountain in Nevada). Specific review of the development of the US regulations is also valuable because the controversial history illustrates well some general points concerning key issues such as:

- the importance of proper allocation of national responsibilities for setting and enforcing radiation standards;
- the potential technical traps which can arise by over-specification of safety requirements (global targets and component requirements);
- the importance of specific points like the form of the standard (dose or risk, individual or collective measures), the timescales to be considered after repository closure, the specifications of future exposed populations;
- the difficulty in defining compliance measures for judging whether a repository will meet the long-term safety criteria;
- the potential for regulatory processes to become very wide, legally complex and painfully slow.

The adversarial legal system in the USA, together with the extreme openness and transparency of the scientific and societal debates make the USA regulatory development process an educational case study for all interested parties. Other nations can learn much from the successful and the failed initiatives in the US debate which has been running intensively since the passing of the Nuclear Waste Policy Act (NWPA) in 1982 (USC, 1982) and is not yet complete.

Who is Responsible for Radiation Protection Standards for Waste Disposal in the USA?

In 1970, responsibility for standard setting in radiation protection in general was transferred to the newly founded Environmental Protection Agency (EPA) from the Atomic Energy Commission (AEC), although the responsibility for implementing and enforcing these standards remained with the latter. Soon afterwards, in 1974, the US Congress responded to the increasing perception of a potential conflict of interest in AEC's roles of promoting and also regulating nuclear power by replacing the AEC with two organisations. The United States Nuclear Regulatory Commission (USNRC) was created for regulating civilian (but not defence) nuclear operations and the Energy Research and Development Administration (ERDA) was allocated the task of promoting peaceful uses of atomic energy and for producing nuclear weapons. In 1977 ERDA became the Department of Energy (DOE).

EPA is charged with setting generally applicable standards for broad classes of activities involving radiation. The class of relevance here is the uranium fuel cycle and, in 1977, EPA promulgated its standard 40 CFR 190 (EPA, 1977) which covered the nuclear power operation part of this, but excluded other aspects including waste disposal. More specific rulings on waste disposal were included in the Nuclear Waste Policy Act (NWPA) passed in 1982.

The NWPA directed EPA to "promulgate generally applicable standards for the protection of the general environment from offsite releases from radioactive materials … in repositories". The Act also directed the USNRC to develop for HLW disposal in mined repositories technical criteria that "are not inconsistent with environmental standards" of the EPA. The generic HLW disposal regulations produced by EPA and USNRC in the years following passage of the NWPA are described below. They are titled respectively 40 CFR 191 (EPA, 1985, 1993) and 10 CFR 60 (USNRC, 1983a). By 1992, developments in US siting programmes for geologic repositories led to new laws relating to repository regulation; the Energy Policy Act (EnPA, 1982) and the WIPP Land Withdrawal Act (LWA) (WIPP, 1992).

From three potential sites selected by a multi-attribute analysis, the US Congress in 1987 (USC, 1987) nominated Yucca Mountain as the primary site for the first repository for HLW and spent fuel. Accordingly, the EnPA directed EPA to promulgate rules for Yucca Mountain. However, in an attempt to break the 10-year log jam of EPA regulations being challenged in the courts, Congress also specified that EPA contract with the National Research Council (the operative arm of the National Academy of Science, NAS) to do a study on standards for Yucca Mountain and specified that the later EPA standards should be "based upon and consistent with the findings of the NAS".

The NAS report appeared in 1995 (NRC, 1995). The technical differences between the NAS findings and the subsequent EPA regulation (EPA, 1999) will be discussed below. The debate was made more complex by the fact that the USNRC had already published its draft regulation 10CFR 63 (USNRC, 1999) and this diverged in places from both the NAS recommendations and the EPA rule. The debate between the different opinions of the NAS group, the USNRC and the USEPA created confusion in technical and public circles. It was resolved by the publication of the EPA final rule (EPA, 2001c) and an amended USNRC regulation (USNRC, 2001).

Meanwhile, the WIPP facility was developed under the EPA general regulation 40 CFR 191 (EPA, 1985), which was drafted in 1985, remanded in 1987 and reinstated in the WIPP Land Withdrawal Act of 1992. Specific criteria implementing and interpreting the 40 CFR 191 rules were produced by EPA in 1998 (EPA, 1998). Under this regulation DOE submitted a compliance certification application (DOE, 1996) for WIPP in 1996, EPA (rather than USNRC which has no regulatory authority for defence installations) certified the facility in 1998 and the first waste shipment to WIPP took place in March 1999.

Technical Criteria in the Early EPA and NRC Regulations

The generic standard 40 CFR 191 set by EPA in 1985 had the following key technical characteristics:

- The primary standard limits the cumulative release of radionuclides to the accessible environment over a period of 10,000 years to values specified in the regulation: this approach, intended to simplify later compliance requirement discussions has been followed in no other country; the limits proposed by EPA were derived from generic calculations indicating that the repository would result in less than 1000 cancer deaths per 10,000 years per 100,000 tonnes of disposed fuel.
- Additional qualitative requirements (e.g. as multiple engineered and natural safety barriers) were intended to provide assurance that standards would be met.
- Additional standards were set for 1000 years on limits to individual members of the public and on groundwater protection.
- The standard was framed probabilistically and it was recognised that "reasonable expectation" of meeting limits was to be demonstrated (rather than "absolute assurance").

When the USNRC published its more detailed compliance criteria in 10 CFR 60 (USNRC, 1983), they attempted to introduce more strongly the concept of "defence in depth" by introducing specific quantitative requirements on each of the major system elements. This was done despite the fact that early drafts of the regulation had produced a vast majority of comments favouring a systems approach. USNRC placed limits on the container lifetime (300–1000 y), the groundwater travel time (1000 y) and the fractional release rates of radionuclides (1 in 10^5/a).

Both the EPA regulations and the USNRC criteria met with major problems following their release (NRC, 1992). Legal challenges to EPA because of international inconsistencies (1000 y versus 10,000 y) and external incompatibilities

Coordination of efforts helps...

with existing deep well injection requirements led to the rules being remanded by the courts until it was reinstated in 1992 in the WIPP LWA. The rule was revised to make all the time frames 10,000 years. The NRC detailed criteria were criticised in the scientific community because they were not properly linked to the system requirements. It was possible to conceive of repository systems where all 3 sub-criteria were fulfilled without meeting the system requirement and, vice versa, it was possible to meet the system standard without fulfilling all 3 sub-criteria.

The NAS weighs in at the request of Congress

Following the directive of Congress in the EnPA that the NAS give guidance, a committee was established, publishing its conclusions in 1995 (NRC, 1995). These diverged in significant ways from the US regulations to date. The 9 key features of the NAS guidance were:

1. An individual risk limit rather than a dose limit was proposed (despite Congress specifying that dose should be used).
2. This limit is for a critical group rather than a maximally exposed individual.
3. The risk level should be set by formalised rule making involving public consultation; appropriate starting figures are the equivalent of 0.1–0.3 mSv/a.

4. The individual risk criteria gives adequate protection for the general public; if a negligible dose of around 0.01 mSv/a can be assumed.
5. There are no scientific reasons for cutting off analyses at 10,000 y; the limit of analysis is set by the geologic stability.
6. Human intrusion should be covered by specifying a stylised intrusion scenario demonstrating that penetration of a single borehole into the repository has insignificant effects on the environment.
7. There is no justification or need for explicit inclusion of an ALARA criterion.
8. Since there is no scientific way of predicting the behaviour patterns of future societies, reference scenarios should be established by rule making.
9. There should be no standards based on best available technologies and there should be no subsystem requirements.

Following the publication of the NAS report, EPA worked for over 3 years to produce a new draft regulation 40 CFR 197 (EPA, 1999). Before this appeared, NRC published its draft regulations for Yucca Mountain, 10 CFR 63 (USNRC, 1999).

The Responses of the Agencies to NAS

Both EPA and NRC proposed for policy reasons to remain with a dose limit. NRC accepted the critical group idea and proposed limiting speculation on the group characteristics by itself specifying assumptions. EPA chose to define rather a "*reasonably maximally exposed individual*" (RMEI), but since they are also rather specific in defining the RMEI, the effects of this difference are minor. The dose levels suggested by EPA and NRC differed (0.25 and 0.15 mSv/a respectively).

In virtually all the other points raised by the NAS Committee, there are no real points of dissention between NAS, EPA and NRC — with one very important exception. This is the time frame for analysis of repository behaviour. Both federal Agencies defend strongly keeping 10,000 years as a time limit for quantitative compliance determination. The NAS arguments (that peak releases are commonly predicted to occur beyond 10,000 years and that the shorter time frame allows almost all safety to be provided by the engineered barriers) are set against the assertion that uncertainties increase with time so that calculations at 100,000 to 1 million years may be meaningless. This issue of timescales is also of wide international interest. For this reason the arguments are gone into below in more detail.

The USA Debate on Timescales

The NAS report stated that "selection of a timescale ... must take into account the scientific basis for the performance assessment itself" and "also involves policy considerations that we have not addressed". The report:

- points out that "one of the major reasons for selecting geologic disposal was to place the wastes in as stable an environment as many scientists consider possible"

- differentiates between changes at the surface and repository performance deep below the ground
- judges that "the timescale for long-term geologic processes at Yucca Mountain is on the order of approximately one million years:"
- concludes that there is "no scientific basis for limiting the analysis" to 10,000 years, and
- recommends applying a risk standard "at times when the peak potential risks might occur".

The NAS recognised that doing full individual risk calculations at long times requires definition of a reference biosphere but believed that meaningful analyses of other parts of the system are feasible beyond 10,000 years. Importantly, it emphasised that peak releases, doses or risks at Yucca Mountain are calculated to arise much beyond 10,000 years and quoted an earlier NAS report which states that a 10,000-year limit "makes compliance rather easy". Lastly, the NAS was of the opinion that uncertainties on the non-biosphere part of the analyses do not necessarily increase rapidly beyond 10,000 years; some uncertainties which can dominate at shorter times (e.g. canister failure rates) can become less important. The NAS also reviewed regulations in other countries, finding that some had no time constraints, some differentiated periods before and after 10,000 years and none had a complete cut off at 10,000 years.

Key Aspects of EPA 40 CFR 197

EPA asked for comments from the public on two approaches to timescale. The first is effectively that proposed by the NAS, the second is application of a quantitative dose limit up to 10,000 years, together with a requirement to "examine disposal system performance after 10,000 years ... to see if dramatic changes in the performance ... could be anticipated". No indication is given of how one judges whether a change is "dramatic". EPA believes that the second approach is preferable. Several categories of arguments (including "policy and technical factors that NAS did not fully address" are given to justify this. These can be summarised as follows:

- consistency with policies for non-radioactive wastes. This is a non-technical issue which NAS did not have to consider. It can be pointed out, however, that some of the EPA requirements for non-radioactive wastes (e.g. "no migration of hazardous constituents for as long as the waste remains hazardous") would indicate, not that compliance periods for radwastes should be shorter, but rather that the effects also of some other hazardous materials should be considered over longer times.
- consistency with other regulations for radioactive wastes, specifically 40 CFR 191 which applies to the same types of wastes as those foreseen for Yucca Mountain. Again this is a non-technical, policy argument not explicitly considered by NAS.

- uncertainties in projecting human exposure over extremely long periods. Whilst EPA "agrees with the NAS conclusion that it is possible to evaluate the performance of the Yucca Mountain disposal system and the lithosphere within certain bounds for relatively long periods", they "believe that NAS might not have fully addressed two aspects of uncertainty". These are the effects of long-term climate changes on choosing a reference maximally exposed individual (RMEI) and the problems in specifying the biosphere conditions in the far future. In fact, NAS considered both these aspects and, precisely for these reasons, recommended using a reference biosphere, defined by the regulator, in order to make analyses feasible. Scoping calculations including effects of climate change on the hydrogeology are possible; the reference biosphere is merely a tool to transform calculated release rates of radionuclides into doses or risks which are to many people more understandable comparisons with other societal risks.
- comparisons with international programmes. EPA states that "many geologic disposal programmes use a 10,000-year regulatory compliance period as a requirement". This period is, indeed mentioned in various national regulations (e.g. Canada, Germany, Finland) but others have no limit specified and specifically all require some consideration of times beyond 10,000 years. Internationally and in Scandinavia, there are proposals for estimating flows or concentrations of radionuclides from the repository in the lithosphere beyond 10,000 years. It is recognised that to judge the acceptability of such estimated results, a comparison basis is needed. This can be derived from back-calculation of potential health effects (which again opens the question of biosphere definition) or from direct comparison with naturally occurring radionuclides.
- "focussing upon a 10,000-year compliance period forces emphasis upon those features over which man can exert some control, such as repository design and engineered barriers". This argument may have some validity in the case of regulations explicitly for Yucca Mountain, since the site is then already defined. As a general argument, however, it is dubious since it may be relatively easy to get compliance through engineered barriers at a site where the degree of natural protection afforded by the geology is lower than could be achieved elsewhere.

Conclusions of NAS on EPA timescale proposal

EPA chose to differ from the recommendation of the NAS, and of the majority of those commenting to EPA on the NAS report, that quantitative analyses of the disposal system should be carried out for compliance purposes far beyond 10,000 years (with a defined reference biosphere) and has retained its earlier recommendation for quantitative compliance assessment only up to 10,000 years. EPA has given a series of policy and technical arguments for this choice and has also included a requirement for consideration of the period beyond. NAS excluded policy considerations from its deliberations. In fact, none of the technical arguments offered by EPA were overlooked by NAS; they do not negate the NAS view that "there is no scientific basis for limiting the time period ... to 10,000 years ...".

However, given that EPA does now require that the performance of the disposal system is examined also after 10,000 years (if the peak dose is calculated to occur then), there may be little difference between the two positions. The major problem may then be that EPA gives no guidance on how analyses should be done for the period of geologic stability beyond 10,000 years and gives no indication of how the results will be considered in judging acceptability. Many calculations for Yucca Mountain already exist, indicating that calculated doses or risks beyond 10,000 years will be very significantly higher than those before this time. To mandate that these results become "part of the public record" but to give no indication of how they will be taken into account seems to postpone rather than solve problems associated with licensing. The NAS suggestion of using the same performance assessment methodology for the engineered barriers and geosphere together with a reference biosphere and judging against the same risk or dose figure is not the only possibility. A less restrictive level could be defined, broader uncertainty limits could be acceptable or, as other programmes have suggested, differing measures can be employed over different time frames. EPA has sought comments on both the NAS suggestion and their preferred alternative, as well as on any other approaches.

In practice the finalised regulations of EPA and of USNRC both require quantitative analyses of potential doses from the repository out to a time of 10,000 years and only qualitative analyses thereafter.

The EPA, NRC Dispute on Dose Limits

The most controversial debate between the two government agencies was on the type and the level of dose limits to be included in their rules. The advice of the NAS was to begin with a proposal such as an individual, all pathways dose limit of 0.1 mSv/a that is internationally common and then to finalise the value by a public consultation, rule-making process. USNRC came out early in favour of a 0.25 mSv/a limit, whereas EPA advocated an all pathways limit of 0.15 mSv/a together with a specific groundwater protection rule equivalent to a maximum dose of 0.04 mSv/a for this particular pathway.

The difference although hotly debated between the Agencies was based more on consistency with existing regulations than on scientific reasoning. Although the EPA staff, like USNRC and NAS, concurred that an individual dose limit can give adequate public protection, they insisted upon including an additional groundwater protection goal. This proposal, which appears to be included mainly for reasons of consistency with other EPA regulations, caused much controversy. USNRC, in its comments on the EPA draft, pointed out that the maximum permissible concentration (MPCs) specified by EPA for some nuclides are equivalent to dose levels far below the suggested global limits.

In the end, EPA's views prevailed because their legal character gives EPA prime responsibility for setting standards. The final rules of both organisations is based on a 0.15 mSv/a limit. The final EPA regulation (EPA, 2001c) and the corresponding amended USNRC rule (USNRC, 2001) have now been published. The at times bitter debate did little to enhance public trust and confidence in either agency and has provided fodder for the legal challenges to be expected.

Appendix 3

List of Acronyms

Acronyms	Meaning
AEC	Atomic Energy Commission of Japan
AECL	Atomic Energy of Canada Ltd.
AIC	active institutional control
AkEnd	Arbeitskreis Auswahlverfahren Endlagerstandorte, Germany (Working Group for Repository Site Selection)
ALARA	as low as reasonably achievable
ALARP	as low as reasonably practicable
ALI	annual limit of intake
BNFL	British Nuclear Fuels plc, UK
Bq	becquerel
CCA	Compliance Certification Application
CFC	Chlorofluorcarbon
CNE	Commission Nationale d'Evaluation relative aux recherches sur la gestion des déchets radioactifs (French National Commission for the Evaluation concerning options over radioactive wastes)
DDL	disaggregated dose-likelihood
DEFRA	Department for Environment, Food and Rural Affairs, London UK
DNA	deoxyribonucleic acid
EBS	Engineered Barrier System
EEZ	exclusive economic zone
EIA	environmental impact assessment
EIS	environmental impact statement
EKRA	Expertengruppe Entsorgungskonzepte für radioaktive Abfälle. (Expert Group on the Disposal Concept of Radioactive Waste)
FAO	Food and Agriculture Organisation

GPA	Global Programme of Action
GWe	Gigawatt (electricity)
Gy	Gray
HEU	highly enriched uranium
HLI	high-level information
HLW	high-level wastes
HSK	Hauptabteilung der Sicherheit der Kernanlagen (Swiss Federal Nuclear Safety Inspectorate)
IAEA	International Atomic Energy Agency
ICM	Integrated Coastal Management
ICRP	International Commission on Radiological Protection, USA (?)
ILI	intermediate level information
IMO	International Maritime Organisation
JNC	Nuclear Cycle Development Institute (Japan)
KASAM	Swedish National Council for Nuclear Waste
KSA	Federal Commission for the Safety of Nuclear Installations (Switzerland)
LL-ILW	long-lived intermediate level waste
LNT	linear no-threshold
MOX	mixed uranium–plutonium oxide fuel
MPBB	maximum permissible body burden
MPC_w	maximum permissible concentration in drinking water
NAS	National Academy of Science
NEA	OECD Nuclear Energy Agency
NGO	non-governmental organisation
NORM	naturally occurring radioactive materials
NRC	National Research Council
NRPB	National Radiological Protection Board (UK)
NWTRB	Nuclear Waste Technical Review Board
OCRWM	Office of Civilian Radioactive Waste Management (of USDOE)
OECD	Organisation for Economic Cooperation and Development, Paris
ONWI	Office of Nuclear Waste Isolation
OSPAR	Oslo and Paris Commission for the Protection of the Marine Environment of the North-East Atlantic
PA	performance assessment
PIC	passive institutional control
PLI	primary level information
QA	quality assurance
RCEP	Royal Commission of Environmental Pollution (UK)
RCRA	Resources Conservation and Recovery Act (USA)
RDP	Repository Development Programme
RMS	record management system
RSK	Reaktorsicherheitskommission (German Commission for Reactor Safety)

SA	safety assessment
SF	spent fuel
SKB	Svensk Kärnbränslehantering AB (Swedish Nuclear Fuel and Waste Management Co.)
SKI	Swedish Nuclear Power Inspectorate
SSI	Swedish Radiation Protection Institute
Sv	Sievert
SZ	saturated zone
UNCED	United Nations Conference on Environment and Development
UNCLOS	United Nations Convention on the Law of the Sea
UNEP	United Nations Environment Programme
USDOE	US Department of Energy (sometimes as DOE)
USEPA	US Environmental Protection Agency (sometimes as EPA)
USNRC	US National Research Council (sometimes as NRC)
UZ	unsaturated zone
WIPP	Waste Isolation Pilot Plant, New Mexico, USA